MAXIMS AND THE MIND

MAXIMS AND THE MIND

Unknowing in the Early Novel
from Bacon to Austen

Kelly Swartz

University of Virginia Press • *Charlottesville and London*

The University of Virginia Press is situated on the traditional lands of the Monacan Nation, and the Commonwealth of Virginia was and is home to many other Indigenous people. We pay our respect to all of them, past and present. We also honor the enslaved African and African American people who built the University of Virginia, and we recognize their descendants. We commit to fostering voices from these communities through our publications and to deepening our collective understanding of their histories and contributions.

University of Virginia Press
© 2025 by the Rector and Visitors of the University of Virginia
All rights reserved
Printed in the United States of America on acid-free paper

First published 2025

9 8 7 6 5 4 3 2 1

LIBRARY OF CONGRESS CATALOGING-IN-PUBLICATION DATA
Names: Swartz, Kelly, author
Title: Maxims and the mind : unknowing in the early novel from Bacon to Austen / Kelly Swartz.
Description: Charlottesville : University of Virginia Press, 2025. | Includes bibliographical references and index.
Identifiers: LCCN 2025022242 (print) | LCCN 2025022243 (ebook) | ISBN 9780813954127 hardback | ISBN 9780813954134 trade paperback | ISBN 9780813954141 ebook
Subjects: LCSH: English fiction—History and criticism | Knowledge, Theory of, in literature | Maxims in literature | Literature and science—England—History | BISAC: LITERARY CRITICISM / Modern / 18th Century | LITERARY CRITICISM / European / General | LCGFT: Literary criticism
Classification: LCC PR830.K54 S93 2025 (print) | LCC PR830.K54 (ebook) | DDC 823/.509384—dc23/eng/20250615
LC record available at https://lccn.loc.gov/2025022242
LC ebook record available at https://lccn.loc.gov/2025022243

Publication of this volume has been supported by the Walker Cowen Memorial Fund.

Cover art: Study for a Portrait of a Seated Gentleman, Allan Ramsay, undated. (Yale Center for British Art, Paul Mellon Collection)
Cover design: Hollis Duncan

For my mother

CONTENTS

	Acknowledgments	ix
	Introduction: Novelistic "Knowledge Broken"	1
1	"Odd Fantastick Maxims": Behn's Partial Knowledge of Love	37
2	The Maxims of Swift's Psychological Fiction	64
3	The New Realism of Literary Generalization in Richardson's *Clarissa*	84
4	Austen's Lessons Not Worth Knowing	103
	Conclusion	125
	Notes	135
	Bibliography	169
	Index	179

ACKNOWLEDGMENTS

In the more than a decade I have worked on this book, I have accrued more debts than I can repay.

I fell in love with the baggy monster that is the eighteenth-century novel while composing terse couplets for an MFA in poetry at the University of California, Irvine. James McMichael was my closest mentor at the time, and I will be forever grateful for his kindness and fierce creative intelligence. At Irvine, they let us poets take one graduate seminar in literature each quarter. When that was not enough for me, some professors kindly permitted me to audit their classes. Douglas Pfeiffer introduced me to Renaissance commonplacing and James Steintrager to libertine literature and philosophy. Their influence can be felt in these pages. Yet it was Jayne Elizabeth Lewis's graduate seminar The Gothic Enlightenment that won my heart. Without her generosity, guidance, and encouragement, I would not have made my way to a PhD program in English literature.

When I arrived at Princeton University, I could not imagine my luck. There I found a community of true friends, unimaginably sharp interpreters, and gorgeous writers. As for dissertation advisors, I had a dream team. Claudia L. Johnson's sharp wit and sophistication served always as the goal at which I aimed in my public persona and my writing. I still repeat her quips and stories to my own students and try to remember to "be deviant" when I write. Sarah Rivett never failed to come to my aid after we first talked Protestantism in a campus café. I want to thank her for her thorough, brilliant responses to drafts and chapters, for our walks through the Institute Woods, and for that work date at Small World when I needed it most. Sophie Gee has always helped me to relax—and laugh—and she read some late-breaking chapters of this book at a critical juncture and provided me with insightful and much-needed feedback.

The dissertation that started it all was supported by a Graduate Prize Fellowship at the Center for Human Values, a Charlotte Elizabeth Procter Honorific Fellowship, and a Cotsen Junior Fellowship, all sponsored by Princeton University.

So many others at Princeton contributed to the dissertation that planted the seeds for this book. I want to thank the many professors in the

English Department with whom I took seminars and talked maxims, especially Diana Fuss, Jeff Dolven, Jeff Nunokawa, Rhodri Lewis, Meredith Martin, Susan Stewart, Esther Schor, Nigel Smith, and Susan Wolfson. It was with members of my graduate cohort with whom I first felt part of a deep intellectual community. Thanks thus go out to Kameron Austin Collins, Eric M. Glover, Anjuli Gunaratne, Francisco Robles, Priyanka Jacob, Emily Vasiliauskas, Mollie Eisenberg, and Rae Gaubinger. Mollie and Rae deserve special thanks for their many years of friendship; both individually and united they have kept me going through personal crises and a pandemic with the help of writing retreats, coffee dates, Zoom writing sessions, and group texts. For many conversations, meals, and late nights, I want to thank additional friends I originally met in graduate school: Henry Cowles, Brandon Menke, Ross Lerner, Liesl Yamaguchi, Amelia Worsley, Sarah Milov, Kyrill Kunakhovich, and Yaron Aronowicz. Finally, no one can boost my confidence quite as much as my dear friend Yanie Fécu.

I could not have completed this book without the support of my wonderful colleagues and students at Adelphi University. My colleagues Katherine Hill, Louise Geddes, Susan Weisser, Igor Webb, and Craig Carson all read and responded to various chapters from the book. Nathan Ross read and commented on an early version of the proposal. A yearly research release from teaching, awarded by my Dean and the Provost, enabled me to complete the book manuscript.

After my first year at Adelphi, I benefited tremendously from an ASECS Fellowship at the William Andrews Clark Memorial Library in Los Angeles, during which time I also participated in the Clark Summer Institute led by Sarah Tindal Kareem and engaged in lively conversation with Helen Deutsch. Thank you to the librarians at the Clark, as well as at Princeton University Library and Swirbul Library at Adelphi, for consistently guiding me through searches and borrowing requests.

Unfortunately, it was only in the final years of this book's composition that I learned the value of consistent fellow readers. It is no exaggeration to say that without my long eighteenth-century writing group—Ula Lukszo Klein, Amanda Louise Johnson, Nicole Garret, and Emily M. N. Kugler—these chapters would be much less cogent and polished. These amazing women helped me see the beauty of eighteenth-century scholarship as a collective and collaborative effort.

I am grateful to two scholarly presses for permitting me to reprint versions of previously published work. Chapter 2 began as an article published in *Eighteenth-Century Fiction* in 2017, entitled "The Maxims of Swift's Psychological Fiction." I want to thank the editor, Eugenia Zuroski, and the two anonymous reviewers whose comments helped me sharpen my argument throughout. Portions of the original article are reprinted with permission from University of Toronto Press. A version of chapter 3 was first published in 2022 as "The New Realism of Literary Generalization in Richardson's *Clarissa*" in *The Eighteenth Century: Theory and Interpretation*. Portions of the original article are reprinted with permission from University of Pennsylvania Press. I am grateful to then-editor Emily Hodgson Anderson and two anonymous reviewers.

I owe a great debt to everyone at the University of Virginia Press, but I am especially grateful to Angie Hogan for shepherding this book through the initial stages of the publication process. The two anonymous reviewers selected by the Press to read the manuscript provided me with invaluable feedback. This book is much stronger thanks to their efforts and suggestions. Thank you as well to Managing Editor Ellen Satrom and to Emily Shelton for her copyediting expertise. Matthew John Phillips ensured that this book's index was sharp and precise.

Shaun, I found you just as I was finishing the dissertation, and then—as always—you listened to me more patiently than anyone else as I sorted through my ideas. Hearing about your own research has deeply altered my understanding and appreciation of scientific practice. The effects of that alteration are everywhere in the following pages. I like to think, too, that you helped me see in Behn how wit and playful inappropriateness can guide a person into a Love without (human) measure. It is no exaggeration to say that I would not have finished this book without you, but the greatest gift you've given me is our family, and yourself. I love you. Nina and Cooper, thank you for letting me love you, and thank you for reminding me every day what really matters in life.

And most of all, thank you to Barbara Swartz, my late mother. You instilled in me a love of books, and I honor you every time I turn a page. This book is dedicated to your memory. It breaks my heart that you can't hold it in your hands.

Introduction
NOVELISTIC "KNOWLEDGE BROKEN"

WHEN DEFOE'S ROBINSON CRUSOE, stranded and alone, runs low on ammunition, he decides to try his hand at domesticating the island's wild goats. His progress is shaky. He remembers capturing an angry, "large old He-Goat" in a baited pit: "I could have kill'd him, but that was not my Business, nor would it answer my End. So I e'en let him out, and he ran away as if he had been frighted out of his Wits: But I had forgot then what I learn'd afterwards, that Hunger will tame a Lyon."[1] This familiar lesson of lion-taming shoos the old goat off the stage, at the expense of its potential for further meaning-making. The goat escapes domestication and flies back to the wilderness, moving in a trajectory opposite to that of Crusoe. Our protagonist reduces the goat-release to its individual human cause: He forgot a truth that we readers will not (should not) forget. He let the goat go, even though he should have known better: Hunger, after all, tames a lion.

Crusoe's maxim seems designed to justify the educational value of his personal narrative, and yet, in relation to himself, the maxim says what he forgot. Crusoe's half-remembered, half-forgotten maxim is not a summit or resting place, but a signpost for a reader (*this way!*). His maxim does not contain the meaning to which it gestures. That meaning is elsewhere.

This book is about how early novelists employed maxims to signal what characters or speakers do not yet understand, cannot confront, or do not know. This not-knowing of a character was in turn a provocation for readers: Abandon your preconceptions regarding how knowledge is passed through experiential narrative before continuing with this fiction. This play of representation and provocation was part of the early novel's engagement with inwardness. Maxims proclaim what is *not* in the mind and the banality of what is. In fiction of the long eighteenth century, maxims are generic, endlessly generatable, and, in a sense, contentless. It is for this reason, strange as it may seem, that Aphra Behn, Jonathan

Swift, Samuel Richardson, and Jane Austen all deployed maxims—pithy, apparently self-evident truths—to represent partial, incomplete, or inadequate understandings of self and world. In relation to a novel reader, this is a profoundly ironic use of the ancient form of the maxim. According to Aristotle, maxims persuaded listeners to accept an orator's arguments by stating preconceptions as truths. Yet, by the early seventeenth century, within the English scientific tradition, Francis Bacon was experimenting with a method by which the maxim could perform a self-canceling rhetorical function. Radically departing from the Aristotelian use, Bacon wrote in collections of disconnected aphorisms as a means of *thwarting* a reader's preconceptions. This method of aphoristic writing could represent, he argued, "knowledge broken," or incomplete knowledge. As a diverse collection of concise generalities without connecting exposition, knowledge broken would lure readers into the comforts of self-evidence only to drop them into the voids between mutually contradictory aphorisms. The method paradoxically guided readers without instructing them, providing a rough shepherding of the human thinker into a material and intellectual encounter with nature's mysteries. Bacon took the method seriously and employed it in some of his scientific writings. Each aphorism he composed (as in the case of the aphoristic *Novum organum* of 1620) pointed his reader in a particular direction. Yet as a collection they diverged, refusing to offer one path to a certain destination.

Jonathan Swift understood this self-canceling rhetorical function of maxims. As a satirist, he exaggerated and amplified the use of the maxim as a sign pointing to an unknown destination. Near the end of *Gulliver's Travels* (1726), we find a deployment of maxims similar to that in *Robinson Crusoe*. A self-satisfied Gulliver details the ingenious ways he has accommodated himself to life among his equine role models: He has found honey to sweeten his water and made leather shoes out of Yahoo (that is, human) skin. Gulliver concludes this report with maxims: "No Man," he writes, "could more verify the Truth of these two Maxims, *That, Nature is very easily satisfied;* and, *That, Necessity is the Mother of Invention.*"[2] Gulliver ironically derives these general lessons for human readers from practices designed to distance himself from human custom (an understatement, given that Gulliver is veering toward cannibalism). The formal consequence is a rhetorical circularity, a black hole that absorbs any light that might usefully be generated by Gulliver's narrative of individual suffering and survival. Gulliver attempts to deliver to us the knowledge he has discovered from experience. But what we see—what we discover—is that he cannot see the loss of his own humanity. One of Swift's points is that fictions of survival may be, at worst, more brutal and, at best, less personally useful to readers than they seem to be.

Maxims—memorable rules or pithy statements of general truth—are everywhere in late seventeenth- and eighteenth-century fiction, which they vex with interruptions, obscurities, and absences. Of course, it can be difficult to assess the maxim's conflictual potential within narrative because it is so commonplace. Pithy citations of Holy Scripture are common in works by Defoe and Richardson. *Don Quixote,* that frequently cited origin-point for the early European novel, plays with the idea of the learning demanded even of fictional histories. When in the prologue Cervantes mourns the absence of such learning in his book, a friend suggests simply sticking in any "few relevant bits of Latin" that he happens to remember.[3] Crusoe's and Gulliver's maxims have no such direct classical pedigree—no traced authorship, as it were—and are not proverbial in the literal sense of that adjective: demonstrative of folksy proverbs, such as Elizabeth Bennet's "Keep your breath to cool your porridge."[4] Instead, Crusoe's and Gulliver's maxims masquerade as self-evident propositions, either derived from individual or collective experience or innately known. As such, they seem to tie new experiences back to the already-known rather than using them as avenues toward future discoveries. And yet, as in the case of Gulliver's brutal, self-exposing maxims, they just as frequently suggest the absence of full and certain knowledge that accompanies experience. In the absence of knowledge from experience, where does that leave readers of early novels?

Histories of the novel in English lean heavily on Enlightenment philosophy to account for the early novel's attempts to be knowledgeable. Crusoe and Gulliver lend authority and credibility to their tales of self-determination by carefully recording observations of nature and society, then rendering in high detail the perceptions and judgments leading up to their narrated actions. Although the above example from *Gulliver's Travels* is satirical, it nonetheless draws heightened attention to certain styles and conventions of an empirical mode. Stylistically speaking, such individual fictional histories seem akin to the natural histories called for by Bacon in the early seventeenth century. In their reporting of observed particulars of experience, Crusoe and Gulliver create an opportunity for "virtual witnessing" for readers: although not directly observing an experiment (or, in the case of realist fiction, first-person experience in the world), readers can nonetheless draw inferences from the reported observations of first-person, or "surrogate," witnesses. John Bender makes this point in *Ends of Enlightenment,* when he connects "surrogate observation in novels by witnesses who stand in for readers" to the "rhetorical formation" of the virtual witness in Robert Boyle's published experiments, which themselves were important to the early Royal Society's constitution of authoritative matters of fact.[5] Here Bender draws the idea of virtual witnessing from Steven Shapin and Simon Schaffer's 1985 *Leviathan and the Air-Pump,* perhaps the most influential work by historians of science among scholars of the eighteenth-century novel.[6] According to Shapin and Schaffer, virtual witnessing contributes to the consensus-building necessary to the creation of scientific fact and the large-scale, public legitimation of modern scientific endeavor.[7] For Bender, the eighteenth-century novel is "an Enlightenment knowledge system that overlapped with those of science and philosophy in a period before the modern disciplines were marked off from one another."[8] Since Bender's work, scholars of eighteenth-century literature and science have continued to engage with the idea of virtual witnessing, albeit from an critical distance. Tita Chico and Kristin Girten both remind us, for example, that such legitimizing witnessing was not only virtual but modest. The perspective from which an experiment was viewed was also to be unbiased and disinterested, and the modest spectator was assumed to be male. Helen Thompson and Girten argue, however, that seventeenth-century experimental philosophers understood the production of knowledge to be more materially relational and immodest, respectively, than Shapin and Schaffer's account in *Leviathan and the Air-Pump* suggests.[9]

This powerful story of the relationship between science and new forms (and pseudoinductive formations) of eighteenth-century narrative fiction with an emphasis on particularized experience has left little room for an assessment of maxims in these novels. The indifference toward maxims among historians and theorists of the novel follows from a belief that the maxim and novel are epistemologically at odds, and that this epistemological conflict has formal foundations. Because of the maxim's apparent self-evidence designed to compel assent, it opposes the rhetoric of virtual witnessing and the open logic of induction, which leaves room for the unbelievable or counterintuitive within scientific discovery. Maxims are apparently *un*novelistic because they are nonparticular, ornamental, and abstract. Novelistic narrative is empirically motivated, but maxims are unempirical.

Such histories of the intersections of the early novel and early science have not, however, considered epistemologies of ignorance as important to the novel's empirical tactics. In his introductory chapter to *Agnotology: The Making and Unmaking of Ignorance*, a 2008 collection edited with Londa Schiebinger, Robert Proctor argues that ignorance has a history that "overlaps in myriad ways with—as it is generated by—*secrecy, stupidity, apathy, censorship, disinformation, faith*, and *forgetfulness*, all of which are science-twitched."[10] Proctor and Schiebinger are hardly alone in their interest in *scientific* ignorance. Since the 1980s, scholars of the sociology of scientific knowledge have studied the construction of ignorance as well as its apparent opposite, knowledge. Reviewing this work, we might say that science needs not-knowing to thrive. Narratives by both Defoe and Swift are alive to this possibility, as demonstrated by the examples I have offered from *Robinson Crusoe* and *Gulliver's Travels*. In both works, the ignorance signaled by maxims aligns with not knowing and partial knowledge rather than with a more willful refusal of knowledge.

Literary critics' omission of epistemologies of ignorance from their accounts of the early novel is strange because Francis Bacon, one of the key figures of the Scientific Revolution in England, insisted that reckoning intimately with our ignorance is one of the first steps toward the unknown. Bacon figures within eighteenth-century studies as the father of Royal Society empiricism and experimentalism. Royal Society members such as Thomas Sprat and Robert Boyle praised him endlessly.[11] Historically, scholars of eighteenth-century literature and science invoke Bacon in discussions of "projectors" because of his belief in our ability to better the human condition through technological innovation.[12] He appears as a traditional inductivist with grand humanistic and imperial tendencies, determined gradually to conquer nature through technoscience. Such work reveals a compelling through-line from Baconian philosophy to Enlightenment improvement (and the colonial violence it could underwrite). Yet in scholarly accounts that gauge the impact of the Scientific Revolution on eighteenth-century novelistic techniques, Bacon is rarely discussed in any detail. This is a mistake, because Bacon had much to say regarding what is distinctive about human thought, a topic long considered central to the novel as a genre. Bacon, who was so many things, was also a theorist of inwardness. As Brad Pasanek argues in *Metaphors of*

Mind, interiority is not an eighteenth-century invention, despite its overwhelming presence within literary scholarship on the period.[13] However, unlike writers of novels, who are often taken as celebrating inwardness, Bacon's interest in the topic was defined by frustration. He was contemptuous of the unaided senses and unaided human intellect and preached humility in philosophical endeavor. In his preface to "The Great Renewal" (*Instauratio magna*), he writes:

> The fabric of the universe, its structure, to the mind observing it, is like a labyrinth, where on all sides the path is so often uncertain, the resemblance of a thing or a sign is deceptive, and the twists and turns of natures are so oblique and intricate. One must travel always through the forests of experience and particular things, in the uncertain light of the senses, which is sometimes shining and sometimes hidden. Moreover those who offer to guide one on the way are also lost in the labyrinth and simply add to the number who have gone astray. In such difficult circumstances, one cannot count on the unaided power of men's judgment; one cannot count on succeeding by chance.[14]

On the topic of intellectual errors, Bacon is wonderfully and insistently eloquent (e.g., "emotion marks and stains the understanding in countless ways which are sometimes impossible to perceive").[15] When in his 1620 *Novum organum* (or "New Organon," which translates to "new instrument"), Bacon communicates his ideas to posterity about his new induction, he does so in a series of antisystematic aphorisms designed to disrupt common patterns of thought, call attention to logical inconsistencies, and make room for gaps—the presence, on the page, of the unknown, or the wilds through which the single mind must travel in order to reach the next resting place—within his program of natural philosophical inquiry. Reproducing the labyrinth on the page might seem strange, given the risks of getting lost. But writing that avoided the labyrinth altogether, creating a world on the page that corresponded in no way to reality, was worse. This technique of using disconnected generalizations to elevate gaps in present knowledge made its way into the early novel, where it coincided with the form's interest in social and moral epistemology, as well as social and moral ignorance. In the early novel, rules of nature that govern individual behavior, for example, may be both commonplace and ungraspable, both known and unknown, as Aphra Behn suggests in her prose fiction. Love, the narrator of *The Fair Jilt* tells us, makes "Ideots . . . wise," "Fools eloquent," and "Cowards brave."[16] The consequences of passion are oxymoronic. As sentence-length expressions of the natural processes of love, they are meaningful *and* nonsensical. They are, we might say, meaning *half*-full.

Reckoning with the early novel's maxims—forms that we can even view as instruments or "assistants" to reason within early science—thus has implications for accounts of the mutuality of literary and scientific methods of learning in the eighteenth-century novel. As a tool for representing the not-known, the eighteenth-century maxim in English was neither a preexisting literary strategy adopted by early scientists nor a scientific strategy adopted by writers of prose fiction. Rather, it was truly a product of coinciding scientific and literary interests, and the early novel's social epistemology was a landscape in which this use of the maxim flourished. This role of empirical unknowing to the novel affected the representation of psychological inwardness, and recognizing this can help us appreciate the degree to which, when it comes to representations of not knowing in early realist fiction, minds are often open, requiring no further penetration by character or reader. Thus, rather than viewing the early novel as a genre devoted to modest, virtual witnessing as a tool for knowledge production, *Maxims and the Mind* takes seriously the not-knowing on display in the novel as part of the genre's commitment to empiricism. At the same time, this book does not reject critical accounts of the early novel as built around a desire for fictional forms of empirical learning and readerly identification. As they engaged with early empirical epistemology and style, early novelists were also learning how they might flex the novel's own growing muscles of insight, including insights into the value of inquisitive unknowing as an attitude to take in relation to human experience and social life.

As microforms within the macroform of prose fiction, maxims became opportunities for writers subtly to question the moral stakes of modern methods of intellectual inquiry, from experimental observation to representations of fictional interiority. Such a maxim appears in a 1711 issue of Swift and Harrison's continuation of *The Tatler* and will serve as another example of how maxims can evoke unknowing in eighteenth-century fictional contexts. Swift's Bickerstaff writes of a propertied man who cursorily reads one of his maid's "Storybooks (as she calls them)" and alights on the "Maxim" that "a Lion would never hurt a true Virgin."[17] The maxim makes an impression, becoming a "sage Observation" that directs Bickerstaff's thoughts after he falls asleep. "I dreamed," he writes, "that by a Law of immemorial Time, a He-Lion was kept in every Parish at the common Charge.... That, before any one of the Fair Sex was married, if she affirmed her self to be a Virgin, she must on her Wedding-Day, and in her Wedding-Clothes, perform the Ceremony of going alone into the Den, and stay an Hour with the Lion let loose, and kept fasting four and twenty Hours on purpose."[18] While the essay initially marks the maxim as silly (positioned as it is within a maid's "Storybook"), Bickerstaff's mind makes it significant: it is the kernel that births a violent fantasy of a publicly witnessed experiment (or "Ceremony") rendering hypocrisy visible. Despite this change in the maxim's status, Swift's essay suggests that both the maid and her employer comfort themselves with the form. Indeed, the maxim is the shared element that keeps both man and woman company in their privacy—the woman while reading, the man while dreaming. The maxim lends the essay its chiastic structure: in reality, this maxim is a fantasy; in the dream world, it becomes a law of nature. Truth and illusion are uncomfortable bedfellows, but bedfellows nonetheless.

Although I have so far spoken only (briefly) of Francis Bacon as the earliest innovator of the modern maxim, this book is interested in multiple seventeenth-century origins: Francis Bacon's aphorisms; François de La Rochefoucauld's *Réflexions ou Sentences et Maximes morales* (*Moral Reflections or Sententiae and Maxims*), first published in French in 1664; and John Locke's maxims, both the innate maxims he rejects in book 1 of *An Essay Concerning Human Understanding* (1689) and the noninnate maxims Locke examines in book 4, which may exist but "can add nothing to the Evidence of certainty of [the Mind's] Knowledge."[19] In some ways, Locke's disdain for the previous epistemological status of the maxim as an innate or grounding axiom of science aligns him with Bacon, whose new induction was determined to reject "current logic," filled as it was with "axioms ... from limited and common experience" that were not challenged or revised to respond to new natural philosophical discoveries.[20] Despite this alignment, Locke and Bacon part ways when it comes to their overall philosophical projects: Bacon's was for an inspirational collaborative program of inquiry, whereas Locke's was an effort to describe the natural history of the individual mind. As a writer, Locke values order. A treatise may be written in fits and starts, but it must be systematized before being presented to a reader.[21] It is only with La Rochefoucauld's *Maximes* that we have an influential return to Bacon's aphoristic "knowledge broken" as a form in which maxims are taken seriously as capable of inspiring inquiry by activating the presence of the not-known. La Rochefoucauld's *Maximes* was a literary-philosophical touchstone for late seventeenth- and early eighteenth-century fiction writers such as Aphra Behn and Jonathan Swift. Gesturing broadly to the antitraditionalism of the new philosophy as well as its interest in apparently imperceptible causes of action and hidden interiors, La Rochefoucauld's maxims reveal Stoic equanimity, or *apatheia*, to be nothing but passionate pridefulness artfully masked. When in 1685 Aphra Behn becomes one of the first writers to translate La Rochefoucauld's work into English as *Seneca Unmasqued, or, Moral Reflections* she places special organizational emphasis on the maxims on love, and she personalizes and genders the universal form of La Rochefoucauld's originals, utilizing a distinctly feminine voice addressing a masculine lover.[22] La Rochefoucauld writes, for example, "The more you love your beloved, the closer you are to hating her," which Behn has her persona Amynta voice as "The more I love *Lysander*, the readier I am to hate him."[23] Behn's rhetorically adept translation of La Rochefoucauld's maxims on love revels in the playfulness of shifting discursive and rhetorical contexts. She treats the *Maximes* as a kind of faux conduct book, and, in playing with conduct prescriptions along with La Rochefoucauld, her translated maxims point toward a potential for the reader to do more with the maximic technique than simply acquire "secret" knowledge about an individual woman's sexual experience. When Swift writes to Pope on reading the *Maximes* just before the publication of *Gulliver's Travels*, he too instigates a play between La Rochefoucauld's universal claims and the particularities of the individual maxim-reader who is also a writer and creator of experiences. Swift portrays the *Travels* as "a Treatis proving the falsity of the Definition *animal rationale*; and to show it should be only *rationis capax*." It is this "great foundation of Misanthropy" upon which "The whole building of my Travells is erected."[24] In a second letter to Pope on the topic of his redefinition of man, Swift refers explicitly to the *Maximes*:

> I tell you after all that I do not hate Mankind. It is vous autres who hate them because you would have them reasonable Animals, and are Angry for being disappointed. I have always rejected that Definition and made another of my own. I am no more angry with [Walpole] Then I was with the Kite that last week flew away with one of my Chickins and yet I was pleas'd when one of my Servants Shot him two days after, This I say,

because you are so hardy as to tell me of your Intentions to write Maxims in Opposition to Rochfoucault who is my Favorite because I found my whole character in him, however I will read him again because it is possible I may have since undergone some alterations.[25]

In response to Pope's "Intentions to write Maxims in Opposition to" La Rochefoucauld (maxims that would, we must imagine, sing of general man's rational virtues), Swift insists on the primacy of the individual view rather than the benefit of general types. For the individual reader, La Rochefoucauld's *Maximes* is a working mirror, quite unlike those satires in which "Beholders do generally discover every body's Face but their Own."[26] Swift's defense of La Rochefoucauld demonstrates, as Behn's *Seneca Unmasqued* does as well, the interpretive activity of personalizing maxims as an act of self-inquiry that leaves room for an ongoing temporal instability of one's "character." These literary engagements suggest that the distance between Bacon's "knowledge broken" and La Rochefoucauld's *Maximes* is not as great as it may seem. Bacon's aphoristic method was designed to inspire the collaborative development of the sciences, while La Rochefoucauld's maxims are more sneeringly pessimistic about the value of human works. Yet, as a form of "knowledge broken" regarding *human* nature and morality, the *Maximes* presents an opportunity for readers to engage with the not-known about themselves (distinct from what we collectively do not know about human or nonhuman nature). Both Bacon and La Rochefoucauld value ignorance-driven inquiry guided by a relationship between reader and literary form.

As the self-canceling maxim makes its way into prose fiction, it becomes a tool for reflecting on the novel's epistemological techniques. Both Behn and Swift, as mentioned, were drawn to the maxim's satiric potential, which is to say they were drawn to the maxim's ability to suggest that an individual's temporally situated perception of the world leads to a uniquely distorted vision. Ironically, for Behn and Swift, ostensible technologies of collective, modest witnessing (experiment, the novel) produced visions no less distorted. Yet Behn saw a way forward through a union of the individual with forces beyond the human. Swift did not. With Samuel Richardson's *Clarissa* (1747–78), maxims become a new way of quantifying fictional history's knowledge.[27] Yet Richardson's incorporation of the maxim into the psychologically realist novel, traditionally seen as elevating private interiority, goes beyond this function of quantification. By weaving literary generalizations from recognizable, contemporary poetry and satire into the thread of his narrative, Richardson reminds his readers of the material reality of maxims on pages outside the fictional world (including the pages of any book of his they may hold). These literary maxims form a bridge more tightly connecting his fictional world to the world of his readers, who are invited to adopt and rearrange Clarissa's maxims alongside Richardson's and their own. Austen, maximic stylist extraordinaire, uses maxims not only to represent a character's ignorance, not-knowing, or inarticulateness, but to continue Richardson's project of hailing a reader, daring her to reject the novel's comfortable truths. Such a rejection would be to her benefit (she would move toward insight) and to her peril (she would move toward insight). Thus, ironically, over the course of the eighteenth century, as we move away from the maxim's contact with early science and anti-Stoicism, maxims in novels come to embody most fully the spirit in which Bacon and La Rochefoucauld intended them to be used. They become reminders that the novel offers only broken knowledge and partial recognition.

From "Knowledge Broken" to Novelistic Unknowing

Maxims and the Mind details a series of surprising conclusions following from its examination of the philosophical origins of maxims we find in early novelistic prose fiction. The first of these conclusions is that eighteenth-century writers of prose fiction used maxims—in collection, and individually—to convey knowledge broken. "Knowledge broken," as defined by Bacon in the context of his new induction, is a mode of arranging ideas wherein the writer presents what she already knows alongside what she does not. As I outline further below, in collections of maxims the space of the not-known or not-yet-known appears materially on the page in the areas unfilled by print. We are not to imagine that knowledge "broken" consists of the shards of some past whole that needs reconstruction (although there is room for this interpretation in Bacon's understanding of the relationship between human natural knowledge and the Fall).[28] Rather, "knowledge broken" is particularized and piecemeal knowledge, even in the case of apparent "general" truth. As piecemeal knowledge, it is incomplete, leaving space for knowledge growth—a space that we should conceive of as temporarily, but not always permanently, empty.

While writers may deploy the maximic *form* of knowledge broken in a variety of ways, when this form interacts with narrative and character, new representational potentials arise. This brings me to the second major claim of this book: I argue that the early novelists who represented inwardness relied in part on maxims to do so. When deployed by characters, maxims can suggest any number of positions on a spectrum of unknowing, from thoughtless inanity to post-traumatic inarticulateness. In each case, these maxims gesture at an inwardness that lacks coherence due to the weighty presence (in the mind and on the page) of the not-known or not-understood. In their contribution to this early novelistic technique, maxims no longer present the gaps in the writer's thinking designed to encourage further inquiry in readers, as Bacon intended. Rather, they present gaps or absences as the content of characters' minds, especially the gaps and absences in the minds of characters who write.

Traditionally, the novelistic emphasis on subjective experience, frequently equated with inwardness, has been traced to Lockean empirical psychology and the construction of the modern subject. Yet this use of Lockean psychology and possessive individualism as a historical framing for novelistic *inwardness* has produced a scholarly blind spot regarding novelistic *ignorance*. Locke insisted on the limits of human understanding of an object's essence. We perceive objects in the world and then mentally categorize and organize them in terms of ideas generalized from previous acts of perception and reflection: knowledge lies with the subject and not with the object.[29] Scholars who trace novelistic epistemology to Lockean empiricism tend to characterize the novel as *the* literary genre that adopts this position. What an examination of early novels' maxims through the Baconian understanding of knowledge broken reveals is that the function and value of novelistic inwardness changes when its key concerns—privacy, originality, self-containment, and self-understanding—are decoupled from the idea that the mind is *either* empty *or* full of knowledge derived from individual intellectual labor exerted upon sense data. Even though the writers in my study use a technique derived from early science to represent the absence of understanding, in the novel this technique leads to a strikingly different picture of inner life than we have been told is created through the genre's empirical investments. The maxims I focus on enabled novelists to represent the absence of unique, meaningful thought derived from experience, either when characters do not have original thoughts, or when they struggle to find their own words for what they have experienced, or when they reject empirical evidence and adopt literary generalization as evidence instead. Despite the apparent imprecision and thus inarticulateness of such maxims, the form affords something unique for readers. Perhaps especially in instances where maxims are ironically deployed within the narrative context itself, readers are emboldened to adopt and adapt them to their own contexts. Maxims enable readers to see themselves in a fictional individual's experiences without foreclosing the possibility that things will not be—and perhaps already are not—the same with them. This leads to a poignancy of partial and limited recognition. It also challenges previous conceptions of inner life as inspired by the early novel. What if we are most true to ourselves when relying on the words of others? What if articulating personal experience through borrowed language is *not* self-protective? What if there is a necessity to publish it despite it being not worth knowing? These are the questions raised by a novelistic inwardness of unknowing.

Consider an example from Samuel Richardson's *Clarissa, or, The History of a Young Lady* (1747–48).[30] As an epistolary novel, *Clarissa* has long been portrayed as particularly loquacious about private thought and feeling. Yet, as Frances Ferguson famously argues in her 1987 essay "Rape and the Rise of the Novel," Richardson understood that the psychological novel was a *social* form, and its status as realistic—as a document of reality—demanded the eradication of uniquely private meaning. Ferguson contends that Richardson responded to this paradox (the self-canceling nature of the psychological novel) by deploying a new "aesthetic": Richardson's "mimesis of distinction." Within this new aesthetic, as Ferguson describes it, "the business of forms"—the act of giving shape to reality by social agreement, from the outside in—presents itself as "unfinished."[31] For example, after Lovelace rapes her, Clarissa wants her experience opened to readers, including the trauma she cannot articulate in matter-of-fact prose. Rather than writing from the first-person perspective, she turns to the maxim. She circulates among her acquaintance her "mad letters," maximic collections of literary generalizations remembered in the immediate aftermath of rape. Paper X, the central document of the "mad letters," is punctuated by blank space. Amid this blankness appear the general truths drawn from unstated yet recognizable, previously published literary sources: "When honour's lost, 'tis a relief to die: / Death's but a sure retreat from infamy" and "For life can never be sincerely blest. / Heaven punishes the *Bad,* and proves the *Best.*"[32] In her interpretation of Paper X, Ferguson focuses on

Richardson's experimental typesetting. The page confronts readers with their previous willingness to accept conventions and read printed pages as if they were handwriting. Ferguson finds this formal confrontation paradigmatic of the unfinishedness of *Clarissa*'s realism. My own interpretation of the piecemeal or unfinished quality of Paper X is different and hinges on its use of maxims. These literary maxims had clear material reality for Richardson's eighteenth-century readers. Just as they had migrated from printed pages into Clarissa's mind and then onto the supposedly handwritten Paper X, they could leap off the page of *Clarissa* and into real-world commonplace books. The gaps and breeches on the page are as central as the literary generalizations to the function of its form for readers. As in the case of Baconian aphorisms, the disconnected literary maxims of *Clarissa*'s Paper X encourage readers to think beyond the page and to let their reality guide them, as prompted uniquely by and in conjunction with Clarissa's papers, toward a new vision. Rather than reveal an interior exposed, the literary generalizations of Paper X are the bases of a reality that Clarissa shares with her readers—that they compose or assemble together. Unlike Baconian aphorizing, which was meant to be restricted to gentlemen philosophers, the disordered Paper X suggests the possible benefits for women readers in circulating and appropriating unowned experience distilled into the form of easily appropriated literary "general truths."

The third and final primary claim of this book is that the maxim sits at the center of a conflict within the early novel itself, one that early novelists wanted to represent to their readers. Many early novelists were drawn in two directions simultaneously: they wanted this new form to operate as an educational tool for readers, especially young women readers. At the same time, they were drawn to characters who struggled to render their own experience therapeutically useful by writing it down. Maxims in the early novel, I argue, often mark the site of this unresolved conflict and gesture at the early novel's role as a modern method of intellectual inquiry whose moral stakes are not always clear for the writers piloting this unique aesthetic project. In other words, ignorance in the sense of the not-known or not-yet-known extends as well to a reader's encounter with imaginative narrative. In maximic moments in novels, readers must reckon with different forms that have historically led to different accounts of the uses of literature (and I mean this in the broadest possible sense, as a knowledge of letters and books) for acquiring knowledge. I started this introduction with an example from Swift's *Gulliver's Travels* that illustrates how the maxim requires a reader to respond to an apparently didactic moment within a narrative in a shockingly unexpected way. Gulliver writes "*That, Nature is very easily satisfied;* and, *That, Necessity is the Mother of Invention.*" The maxims, with a self-assured closure only achievable by absolute banality, contribute to the predictable structure of individual triumph into which Gulliver seeks to press his experience. Yet the full passage leading up to Gulliver's familiar maxims is as follows: "I soaled my Shoes with Wood which I cut from a Tree, and fitted to the upper Leather, and when this was worn out, I supplied it with the Skins of *Yahoos,* dried in the Sun. I often got Honey out of hollow Trees, which I mingled with Water, or eat it with my Bread. No Man could more verify the Truth of these two Maxims, *That, Nature is very easily satisfied;* and, *That, Necessity is the Mother of Invention.*"[33] Gulliver's account of his life in Houyhnhnmland conveys his experience in the form of retrospective narrative. He presents his maxims as part of this endeavor—as knowledge distilled, if not induced, from his experiential report. Yet, despite our familiarity with the maxims, a familiarity that might encourage us to view Gulliver's deployment of them as straightforwardly rhetorical, we cannot blithely nod along as they cap his smug report of ingenious resource management: he has satisfied his "necessities" quite literally with the bodies of other people. Although Gulliver ostensibly writes a "real" history grounded in wisdom hard-won from individual experience, the effect is to portray loss: of home, of humanity, *and* of comforting rhetorical banalities. Those maxims that were once so comforting become the site of comfort's absence, an unknowable emptiness at the heart of an experience meant to be valuable to others. Gulliver's maxims, as forms with a history in rhetoric and in science, provoke a reader to revisit the source and purpose of fictional personal narrative and to see it two ways at once. On the one hand, Gulliver's narrative is intended to be self-consciously educational for a reader; on the other, the act of writing for Gulliver is an individual attempt at recovery and recognition. Can the two purposes exist simultaneously, or must they undermine one another? This is a question of genre posed to the reader—a question central to the early novel as I present it.

According to Marthe Robert, the novel was a literary "upstart" that "[graduated] from a discredited sub-category to an almost unprecedented Power" by the twentieth century: "With the freedom of a conqueror who knows no law other than that of his unlimited expansion, the novel has abolished every literary caste and traditional form and appropriates all modes of expression, exploiting unchallenged whichever method it chooses."[34] For Robert, the novel "knows neither rule nor restraint," yet the eighteenth-century novel did know—and embrace—rule in the form of the maxim.[35] Yet, as the following pages will show, this embrace did not lead to retraction or restraint. The maxim's conflictual encounter with narrative creates an epistemological (or agnotological) interruption that fuels questions about what and how fiction can know about us and our world.

My choice to utilize "maxim" (as opposed to "aphorism") to name the primary object of analysis in what follows, and to range Bacon's examples under that name, has a heuristic value not only for eighteenth-century studies and the history and theory of the novel in English, but also for literary studies more broadly. "Maxim" was the dominant term for this form—both in isolated instances and in collection—in literary works and contexts throughout the eighteenth century. Both Ephraim Chambers's *Cyclopædia* (1728) and Johnson's *Dictionary of the English Language* (1755) define "maxim" and "aphorism" synonymously. In Chambers's *Cyclopædia,* "aphorism" is "a Maxim, general Rule, or Principle of a Science; or a brief Sentence, comprehending a great deal of Matter in a few Words."[36] Johnson defines "aphorism" as "a maxim; a precept contracted in a short sentence; an unconnected position."[37] Yet "maxim," as it has fallen out of common usage, has become associated primarily with morality and doctrine, whereas "aphorism" has retained the prestige of both literary and philosophical associations. My study aims to expand our sense of the literary and philosophical richness of maximic usage that we now see as bland and dogmatic.

From the "Plain Style" to Form as Method in Histories of the Empirical Novel

Historians and theorists of the novel have largely been uninterested in maxims, despite their ubiquity in the narrative fiction of the long eighteenth century. Early novelists, we are told, used their craft to create a world constituted by plausible fictional particulars in order to advance a convincing impression of individual life.[38] Because these particulars and their arrangement were invented, the novel as a whole could lend a coherence to life that individual experience lacked.[39] This "formally realist" inventiveness of eighteenth-century British fiction depended in part on novelists' adoption and transformation of principles and styles of empiricism. Prose fiction, the story goes, increasingly came to report on its narrative events in a "plain style"—a focus on using words to denote real, material things, without ornament. Just as natural philosophers used the plain style to refuse the misleading constructions of rhetoric in their accounts of natural phenomena, so fictionists in this new empirical mode deployed the plain style to avoid misleading constructions of human life. In histories of the novel's rise, the role of the plain style has come to be deeply interwoven with the question of whether this form of novelistic mimesis could extend human perception, enabling readers vicariously to have experiences on par with those created by the mental impressions of objects encountered in everyday life. In some theories of the early novel, this adoption of the plain style was further justified by the argument that it would enable readers—especially young, inexperienced women readers—to reason more soundly about the real-world dangers they faced. Of course, empiricism's influence on narrative does not stop with the plain style discussed above. Fictionists reproduced sensory experience and perceptual reality through more idiosyncratic uses of literary form. By manipulating perspective and causality through literary means, writers could defamiliarize common experience in order to solicit wonder in a reader, a technique we can trace back to the writings of seventeenth-century Royal Society empiricists such as Robert Hooke.[40] As Roger Maioli demonstrates in his work on early accounts of the novel's empiricist underpinnings, this stage of empirical influence was central to the early novel's development, and it is also part of a larger modern story about the perceived relationship (and differences) between the knowledge produced by imaginative works and knowledge as produced by the sciences.[41]

Yet, by focusing on how the early novel formally addresses the fraught relationship between sensory perception and representation, historians of literature and science have overemphasized the role of mimesis in the novel's epistemological aims. As Aaron Hanlon argues, we scholars of the early novel have "largely confined ourselves to addressing the relationship between fiction and knowledge through the back door of novelistic realism, reducing epistemic matters to matters of form and representation."[42] I agree with Hanlon's call for scholars of empirical knowledge in the novel to move away from the heretofore prevailing focus on problematics of what we might call "classical novelistic realism." A focus on mimesis brings certain things to the fore and relegates others to the background. Within such a framework, what emerges as primary is the relationship between a perceiving subject and the objects perceived, or between subjective experience and reality. Pushed to the background are models of rational inference, as Hanlon argues, but also ignored is the importance within scientific practices and procedures, since Bacon's time, of acknowledging what we do not yet know and recording margins of error. Literary form can play a role in this as well, and letting it occupy that role forces us to stop looking for literature to critique science and to start paying attention to early accounts of how science aimed to critique itself. Early novelists were interested in this mode of scientific self-critique, and they built into the novel their own distinctive methods of novelistic self-critique. Both of these self-critiques—science's and the early novel's—begin from the foundational assumption of the human mind as part of nature and the products of the human mind as artifacts of that relationship.

Despite their differences, theorists of the early novel such as Bender, Maioli, and Hanlon treat science as a means to banish ignorance, especially the ignorance of tradition. Their accounts of eighteenth-century literature, especially the novel, take for granted a view of science—to use Peter Wehling's terms—as "the eminent modern institution that produces knowledge and, at the same time, *reduces* or even *eliminates* pre-existing ignorance."[43] In his history of the development and presence of the concept of "scientific ignorance" in the sociology, history, and philosophy of science since the twentieth century, Wehling explains that when science is viewed in this way, "ignorance, as that which is merely 'not-yet' known, figures only in the very first stages of scientific research and, therefore, is usually held to be an ephemeral phenomenon, not worthy of closer inspection."[44] When we hold eighteenth-century literature to this standard, insisting that for it to be modern it must eliminate ignorance, either the ignorance of traditional philosophy *or* the ignorance of science— we prevent ourselves from observing a deep object of fascination for eighteenth-century writers of prose fiction, especially those compelled by representations of the not-known. I will turn now to a deeper look at Francis Bacon's interest in this regard.

Maxims in Baconian Science

In England, the maxim's modern evolution begins with Francis Bacon. By the early seventeenth century, the relationship between instruction and invention within natural philosophy had become complicated. Bacon developed his ideas regarding the weakness of traditional methods of "inventing" (or discovering) knowledge alongside his reading of early modern reformers of dialectic, the latter being itself an offshoot of Scholastic demonstrative logic. Scholastic logic derived from Aristotelian formal logic. Truth in the context of formal logic requires the demonstration of the necessary connection between propositions and the terms within those propositions. It is the study of valid inference. To Bacon, the problem with Scholastic demonstration and early modern dialectic was that they both asked for belief and not for questioning. Such proofs were overly instructive, leaving no room for altering or extending past conclusions. Logic, a human technology for rational thought and the invention of knowledge, had been corrupted by dogma. Bacon believed that Aristotle had dominated the great minds of previous generations and blamed him for the absence of respectable scientific inquiry into the natural world. While knowledge in some arenas (such as moral philosophy) could still be "invented" using traditional logical proofs, natural philosophical knowledge could not hope to advance under such conditions. New logical methods of causal inference were needed. Instead of "anticipations of nature" we needed "interpretations." Bacon's new induction, which he called the "interpretation of nature," is a mode of "reasoning which is elicited from things in proper ways" by "elicit[ing] axioms from sense and particulars, rising in a gradual and unbroken ascent to arrive at last at the most general axioms."[45] Throughout this process, described by Bacon in the *Novum organum*, reason needs "assistants to the senses," and assistance "not so much from instruments as with experiments . . . devised and applied specifically for the question under investigation with skill and good technique."[46]

We saw in the previous section that most scholarship discussing the relationship between empiricism and the early novel finds that the strain of eighteenth-century fiction we now call "realist" or "protorealist" was designed to function as a technology of perceptual augmentation along the lines of a microscope or telescope. Furthermore, this literary technology (Shapin and Schaffer trace it back to Robert Boyle's "naked way of writing") was seen as a more faithful and modest record of experience because it abandoned any rhetorical techniques of persuasion.[47] In its accounts of early modern science, such scholarship emphasizes the importance of experiment and observation within the Royal Society. It is a commonplace within this scholarship that experiment and observation were practices designed to ensure a philosophical focus on *things* over *words*. As G. A. J. Rogers writes of Bacon's influence on the Royal Society:

> This spirit [of experimental inquiry pursued by Boyle and Locke] was essentially the same as that whose great spokesman was Francis Bacon, the intellectual forerunner of the Royal Society.... Central to this was the belief that careful observation and experiment were much more important than theory to a correct account of the natural world. Of course, there was a place for hypotheses, but any such hypotheses must be rigorously tested against the world. When the Royal Society was formed in 1662, it took as its motto "*Nullius in Verba*"—nothing in words. In so doing it committed itself to the notion that knowledge of the natural world was to be obtained not by verbal exchanges but by careful empirical enquiry.[48]

By Rogers's account, Locke follows Boyle who follows Bacon in warning against the contaminating role of words in natural philosophical knowledge.

The story of Baconian empiricism as a battle for things over words discussed above misses another important element of scientific thinking that affected the novel: method, and not only induction—the method for the interpretation of nature—but aphorisms, the method of writing best suited to transmit that ongoing interpretation in partially completed form. Bacon certainly warned of the dangers of verbal exchange to natural philosophical inquiry when he described the "*idols of the marketplace*" and "*idols of the theatre*," two of the four "idols" or "illusions" that, he argues, "got a hold on men's intellects in the past."[49] Of the idols of the marketplace Bacon writes, "Men associate through talk; and words are chosen to suit the understanding of the common people.... Plainly words do violence to the understanding, and confuse everything."[50] Yet Bacon recognized not only the need for words, but also that science is not simply empirical observations; it is also logic, rhetoric, and creating theories based on existing knowledge. There is a need for words in order "to provide more reliable and secure directions for present and future generations."[51] Indeed, the full title of the Latin *Novum organum* translated into English is, "The New Instrument, or True Directions for the Interpretation of Nature." Bacon does not discuss scientific theory as we understand it today—as originating in hypotheses tested and supported by numerous empirical investigations. Bacon does, however, theorize about method—both methods for inventing knowledge *and* methods for transmitting knowledge to others—when he advances a set of principles about how best to build the new natural philosophy.

Bacon's inductive process may sound familiar, but method is more complicated for Bacon than Rogers or others allow. Bacon's views on methods of transmitting partial knowledge were informed by his beliefs about the minds of readers. His accounts of readerly minds are richly metaphorical, as one would expect from one of our greatest humanist essayists. In a famously vitriolic passage from book 1 of his 1605 *Of the Proficience and Aduancement of Learning, Diuine and Humane* (hereafter *The Advancement of Learning* or the *Advancement*), the name Aristotle appears only in passing, but it is closely associated with the "vain altercations" and arachnid intellects Bacon attributes to Scholastic theologians. Bacon attacks the Schoolmen (or Scholastics) as purveyors of knowledge, but the problem begins with their own reading. The "schoolmen," Bacon writes, had "sharp and strong wits, and abundance of leisure," but

> their wits being shut up in the cells of a few authors (chiefly Aristotle their dictator) as their persons were shut up in the cells of monasteries and colleges; and knowing little history, either of nature or time; did out of no great quantity of matter, and infinite agitation of wit, spin out unto us those laborious webs of learning which are extant in their books. For the wit and mind of man, if it work upon matter, which is the contemplation of the creatures of God, worketh according to the stuff, and is limited thereby; but if it work upon itself, as the spider worketh his web, then it is endless, and brings forth indeed cobwebs of learning, admirable for the fineness of thread and work, but of no substance or profit.[52]

In this passage the Schoolmen's minds and bodies are equally confined, their "wits" and "persons" imprisoned, in thrall to Aristotle "their dictator." This absence of liberty prevents their access to "matter," such that the systems they build end up revealing more about themselves than about the world's actual workings. Scarcity of matter does not prevent them from bringing their intellects to bear on the scraps: Their "wits" spin out silken garments threadbare even before use. And who thrives under such conditions, but spiders, happy to belabor their own peculiar material in the dark? This figurative flight lands back at literal books—in this case, merely containers of cobwebs, or empty controversy. The "absolute resignation or perpetual captivity" of Scholastic disciples directly causes the "laborious webs of learning" that fill their books (*A* 144). Readers of the Schoolmen's "webs" encounter products of boredom and confinement instead of reflections on and directions for encounters with the actual material world.

Bacon's disgust with the Schoolmen's submission to Aristotle echoes through his similar abhorrence of traditional philosophical *methods* of logical demonstration when it comes to seeking knowledge about the natural world. Logical demonstration involved inferring conclusions from premises according to long-established rules of deductive reasoning. Consider syllogism as a prime example. Bacon, however, believed that new rational methods were needed for the invention of natural knowledges. "Method" became a key word in Bacon's development of new procedures of logical proof that would remake classic induction.[53] Furthermore, these new and different methods of knowledge creation would require different forms of writing to communicate both the methods themselves and the knowledge they produced. Through his theorization of aphorizing as a method of writing to communicate such new methods of induction, Bacon responded in part to questions many in early modern Europe were asking about how something should be made known.

Bacon's clearest account of "Method" occurs in the *Advancement*. There he partitions human learning into three branches, each corresponding to the faculty of man's understanding that produces it: "History to his Memory, Poesy to his Imagination, and Philosophy to his Reason" (175). Bacon places "knowledges as do handle and inquire of the faculty of Reason" under the rubric of logic—the "Art of Arts"—and subdivides them into "the Arts Intellectual," which are four in number, divided "according to [their] ends": "the Art of Inquiry or Invention; Art of Examination or Judgment; Art of Custody or Memory; and Art of Elocution or Tradition" (219). The "Art of Elocution or Tradition" Bacon further subdivides into three parts: "the organ of tradition" (speech or writing), "the *method* of tradition" (mode of presentation—and I venture anachronistically to suggest the modern term "form"), and "the illustration of tradition" (rhetoric) (230, emphasis mine).[54] This tour of the arts intellectual is important for understanding Bacon's conception of the role of *discovering* arts and sciences as opposed to merely *teaching* them.[55] The "Art of Inquiry or Invention" is the art of discovering knowledge. But the *method* of writing, otherwise known as the mode in which that discovery would be presented to others, played an important role in the ongoing progression of knowledge for Bacon.[56]

Bacon distinguishes between the invention or discovery of "Arts and Sciences" and the invention or discovery of "Speech and Arguments" in the following way: Rhetorical methods were suitable for the invention of the latter, but for the former, the rules of invention, judgment, and delivery must be one and the same.[57] But such a process—Bacon calls it true induction, or the "Interpretation of Nature"—was not yet in existence and would take much work and collaborative effort to achieve. It is within this context that Bacon introduces his fuller discussion of "method." In Bacon's approach to the topic, "syllogism" plays a large role, because it was synonymous with logic—the rules of rational inference—at the time:

> Method hath been placed, and that not amiss, in Logic, as a part of Judgment: for as the doctrine of Syllogisms comprehendeth the rules of judgment upon that which is invented, so the doctrine of Method containeth the rules of judgment upon that which is to be delivered; for judgment precedeth Delivery, as it followeth Invention. Neither is the method or the nature of the tradition material only to the use of knowledge, but likewise to the progression of knowledge: for since the labour and life of one man cannot attain to perfection of knowledge, the wisdom of the Tradition is that which inspireth the felicity of continuance and proceeding. And therefore the most real diversity of method is of method referred to Use, and method referred to Progression; whereof the one may be termed Magistral, and the other of Probation. (*A* 233)

Bacon states here that rules of judgment—the rational arts—act not only on what is invented. "Method" also refers to rules of judgment that guide the presentation of material already invented. These methodical rules of judgment concerning delivery can be divided into two types: "Magistral" and "of Probation." Magistral methods are best for a knowledge that is to be of immediate, practical use ("method referred to Use"). Such knowledge is meant to be absorbed and acted on without question. Methods "of Probation" are more suitable for a knowledge undergoing development ("method referred to Progression"). These methods of probation are designed to "inspire" protracted examination and the gradual advancement of knowledge.[58]

We should pause over that term "probation." Bacon has in mind here "the action or process of putting something to the test; trial, experimental; investigation, examination."[59] In *De augmentis scientiarum* (1623), Bacon's Latin translation and expansion of the *Advancement*, method of probation is called "initiative," "as in ceremonies admitting a newcomer to a group."⁶⁰ In his 1674 English translation of *De augmentis,* Gilbert Wats translates the relevant passage on methods of probation as follows:

> Wherefore let the first difference of *Method* be set down, to be either *Magistral,* or *Initiative:* neither do we so understand the word *Initiative,* as if *this* should lay the ground-work, the *other* raise the perfect building of *Sciences;* but in a far different sense, (borrowing the word from sacred Ceremonies) we call that *Initiative Method,* which discloseth and unvails the Mysteries of Knowledges: For *Magistral teacheth, Initiative insinnuateth: Magistral* requires our belief to what is delivered, but *Initiative* that it may rather be submitted to examination. The *one* delivers popular *Sciences* fit for Learners; the *other, Sciences* as to the *Sons of Science:* In sum, the one is referred to the use of *Sciences* as they now are; the other to their continuation, and further propagation.⁶¹

Methods of probation and methods of initiation are one and the same, yet the terminological shift from the former to the latter has an important consequence for the way knowledge is conceived. Knowledge under methods of probation is imagined as incomplete and in need of progression, whereas knowledge under methods of initiation is imagined as arcane and secret. The receiver of initiative knowledge is invited, as it were, to move through ceremonial rites that will induct him into the "Mysteries of Knowledges." The initiative method is the probative method, for it still encourages examination, but the term "initiation" suggests a more exclusive collective of examiners.

Although Bacon's initial treatment of magistral and probative methods is somewhat balanced in the *Advancement,* what immediately follows shows that Bacon views the magistral as occasionally necessary but ultimately inferior. This is because knowledge delivery, according to Bacon, is particularly susceptible to the corruptions of "a contract of error between the deliverer and the receiver: for he that delivereth knowledge desireth to deliver it in such form as may be best believed, and not as may be best examined; and he that receiveth knowledge desireth rather present satisfaction than expectant inquiry; and so rather not to doubt than not to err: glory making the author not to lay open his weakness, and sloth making the disciple not to know his strength" (233). Instead, knowledge should be delivered "as a thread to be spun on . . . in the same method wherein it was invented, and so is it possible of knowledge induced" (233–34). As mentioned previously, Bacon is fond of metaphor, and this one deserves unpacking, though it is notoriously opaque. By referencing the art of hand-spinning, or pulling and spinning thread from a raw material such as wool or flax, Bacon figures knowledge making as artistic making. Such thread is spun continuously by the artisan who examines its thickness carefully, making necessary adjustments while working. If the thread becomes too thin, additional wool is pulled in while spinning continues. Knowledge "delivered as a thread to be spun on" is partial knowledge delivered continuously, extended without break, because it is delivered in a method enabling the receiver to incorporate it—albeit with tactile effort and art—into existing understanding that will itself be useful material for future art. We might compare this process of ongoing knowledge spinning with the metaphor of the Schoolmen's cobweb treatises mentioned above. The Scholastic "wits" work alone in a confined space, with only their own materials. Their need for and lack of stimulation from external objects results in overly intricate and ephemeral "webs." Bacon's homespun and ever partial but always extendable knowledge is more sustaining than arachnid majesty.

The spinning metaphor is not the only one Bacon uses in his depiction of methods of knowledge delivery that encourage ongoing growth. Another figure imagines minds as providing soil for sustainable cultivation. In an effort further to explain how, under probative methods, rational rules of invention double as rules of delivery, Bacon turns to horticulture. Although we may not distinctly remember how we invented or discovered the knowledge we possess, Bacon writes, "a man may revisit and descend unto the foundations of his knowledge and consent; and so transplant it into another as it grew in his own mind" (*A* 234). Here Bacon suggests that delivering natural knowledge ought not to be akin to delivering only the fruits of one person's individual labors for consumption by the receiver. Instead, the receiver of knowledge adopts the plant, roots and all, and thus is responsible for tending it.

Metaphors are conceptually useful, but what does a probative method look like on the page, one of the vehicles for knowledge transmission? Since Bacon declared probative methods a desideratum, we have only his brief comments to help us understand what he envisioned.[62] Bacon was clear on this point, however: "The delivery of knowledge in Aphorisms" is one such probative method. When aphorisms appear in the discussion of method in book 2 of the *Advancement*, Bacon somewhat confusingly contrasts them with "Methods" (i.e., systematic expositions).[63] "We may observe that it hath been too much taken into custom," writes Bacon, "out of a few Axioms or observations upon any subject to make a solemn and formal art; filling it with some discourses, and illustrating it with examples, and digesting it into a sensible Method; but the writing in Aphorisms hath many excellent virtues, whereto the writing in Method doth not approach":

> For first it trieth the writer, whether he be superficial or solid: for Aphorisms, except they should be ridiculous, cannot be made but of the pith and heart of sciences; for discourse of illustration is cut off; recitals of examples are cut off; discourse of connexion and order is cut off; descriptions of practice are cut off; so there remaineth nothing to fill the Aphorisms but some good quantity of observation: and therefore no man can suffice, nor in reason will attempt to write Aphorisms, but he that is sound and grounded.... Methods are more fit to win consent or belief, but less fit to point to action; for they carry a kind of demonstration in orb or circle, one part illuminating the other, and therefore satisfy; but particulars, being dispersed, do best agree with dispersed directions. And lastly, Aphorisms, representing a knowledge broken, do invite men to enquire farther; whereas Methods, carrying the shew of a total, do secure men, as if they were at furthest. (234–35)

Aphoristic writing conveys only "pith and heart" without unnecessary discursive filler. As a form it draws on the physicality of the page, demonstrating incompleteness through blank spaces between entries. Aphorisms are antisystematic, because for Bacon the systematically organized treatise neither "lays open" the writer's weakness nor permits the reader to know her strength in action, following the "dispersed directions" through a world of "particulars." The whole point of a probative method is not to display the writer's mastery but rather to promote the reader's thought while suspending judgment, a state of mind useful for ongoing scientific inquiry.[64] The reader then provisionally accepts the current state of partial knowledge in preparation for continued spinning (to extend Bacon's textile metaphor) or ongoing growth (to borrow his transplantation figure). In this way, aphorisms trouble the once smooth relationship, or "contract of error," between writer and reader. They do not satisfy the writer's vanity and enable the reader's sloth as more systematically organized treatises are wont to do. Bacon's point is that if something immediately strikes us as true we should be suspicious, and so aphorisms are designed to be substantial but difficult to comprehend. Antisystematic aphorizing provokes investigation: "Aphorisms, representing a knowledge broken, do invite men to enquire farther."[65]

The content of the aphorisms does not "represent a knowledge broken." Their form does. As a seventeenth-century prose stylist, Bacon operated within a literary tradition in which an author's skill was determined by—as Ian Watt says of prenovelistic imaginative writers—"the literary sensitivity with which his style reflected the linguistic decorum appropriate to its subject."[66] The writer bestowed "extrinsic beauties . . . upon description and action by the use of rhetoric."[67] (Nowhere does one see these rhetorical flourishes more than in Bacon's *Essays*.) We can appreciate the innovation of Bacon's use of aphoristic form within this literary context. Despite his connection to traditional prose style, Bacon was willing to invent and innovate with form. His use of aphorizing is rhetorical in that it has designs on a reader, but he is not simply playing at rhetoric by hanging "extrinsic beauties" onto otherwise substantive knowledge. According to Bacon, the technology of writing—in all of its material aspects, including genre and form—could be manipulated by men on behalf of the great renewal of learning. This is what he attempts to do with the aphoristic method of writing.

The aphoristic method of writing does offer a large degree of liberty to readers that Bacon tempered by filling his aphorisms with mockery of the flaws of human behavior and thought. For example, in his *Novum organum*, Bacon presents his new scientific method—induction—entirely in aphorisms. The work is in two books, with the first dealing extensively with the famous "Idols" of the mind and the previous causes of errors in the sciences. The work was to replace Aristotle's flawed *Organon* and to outline, for the first time, the true method for discovery in the sciences.[68] Many of the insights of his early aphorisms in book 1 of the *Novum organum* come from the sense that they humble us, asking us to see ourselves as fallen creatures with a need to repair our equally fallen relationship to the natural world and the knowledge it provides. When, in book 2 of the *Novum organum*, Bacon lays out procedures for conducting the Interpretation of Nature (his new induction), many of the aphorisms expand to more than several pages, some including organizational tables and extended examples of practice.[69] The below excerpt is taken from a 1733 English translation of book 1:

APHORISM IX.

9. The Root of all the Mischief in the Sciences, is this; that *falsly magnifying and admiring the Powers of the Mind, we seek not its real Helps.*

APHORISM X.

10. The Subtilty of Nature, far exceeds the Subtilty of the Sense and Understanding; so that the sublime *Meditations, Speculations,* and *Reasonings* of Men, are but a kind of Madness; if a fit Person were to observe them.

APHORISM XI.

11. As the *Sciences* now in being, are useless in the Discovery of Works; so is the present Logic in the Discovery of the *Sciences*.[70]

Aphorism follows aphorism follows aphorism. Unordered segments, they beg to be reconciled with one another, yet their true glory is in their inability to be processed in any simple, single way. As we can see, here Bacon breaks down the foundations of traditional scientific demonstration to make way for his new Interpretation of Nature. This new method is probative: In its experimental methods as well as its mode of typesetting and printing it will convey knowledge as "a thread to be spun on." Later in the *Novum organum* Bacon details more concrete techniques and methods by which philosophers might gather and order observations such that future thinkers may continue the inductive work of putting them to the test. Even here, however, in the early aphorisms of book 1, Bacon experiments with probative methods of presentation. Knowledge delivered is partial; the receiver is equally reminded of its absences, or her ongoing ignorance. As knowledge is made, so is nonknowledge. The opacity of the aphorisms as well as the empty gaps among them are reminders—intellectual versions of the felt imperfections in thread—of the earthiness of the knowledge making process. The key to induction is following Nature's lead by abandoning our insane reliance on the "naked intellect."[71] Just as the hand needs tools, the mind needs constraints and instruments to guide it, and one such instrument is this formal reminder of nonknowledge. He writes: "Human knowledge and human power come to the same thing, because ignorance of cause frustrates effect. For Nature is conquered only by obedience; and that which in thought is a cause, is like a rule in practice."[72] For Bacon, the goal of following Nature's paths instead of dallying behind "the mind's endless and aimless activity" was a strategy to regain human control of nature.[73] Bacon calls for men to follow his plan for constructing "a new and certain road for the mind from the actual perceptions of the senses" and thus by this road "to penetrate further; and not to defeat an opponent in argument but to conquer nature by action."[74] We must obey nature before we can become its master. Aphorisms would help in this endeavor toward ultimately controlling nature for human purposes, creating a sort of literary-formal screen through which natural philosophers would view nature, correcting for the distortions of their intellect and helping them to identify and examine the unknown.

Interiors and Anti-Interiors from Bacon to the Early Novel

This history of the maxim helps explain early novelists' use of the form to represent the not-known. Understanding this history thus obliges us to revisit accounts of novelistic interiority and early novelistic realism. As discussed in the second section of this introduction, Locke's empirical psychology, which focused on the limits of human understanding with regard to objects, led to a greater emphasis on our epistemological grounding in sensory experience *and* our internal reflections. As Locke explained, empirical knowledge of the natural world could never approach certainty, only probability. Yet the role the individual mind played in the personal acquisition of empirical knowledge, however probabilistic, had a further benefit: the consolidation of the subject through knowledgeable self-reflection and her liberation from the world of things. This individual owner of sensory knowledge became the model for the construction of realistic fictional characters. Following from this logic, recent reviews of theories of the novel and the modern subject—such as that found in Sandra Macpherson's *Harm's Way*—claim that "life—an existence freed, through the agency of thought, from a purely material causality—is what novels want, and what they render."[75] Macpherson charts how Georg Lukács and Franco Moretti link the novel's emphasis on freedom to both "interiority" and "subjective possibility."[76] According to such arguments by Lukács, Moretti, and others, this is the story of the birth not only of the modern subject, but also of the liberal subject who will reign absolutely over her life and the meaning she makes of it.

The Baconian account of the mind's position in relation to natural knowledge is much more rigorously deflating. There is certainly neither freedom nor autonomy in the mind's webs, only servility. This account of Bacon's "deflating" sense of subjectivity might seem to conflict with more traditional accounts of his confidence in his new induction. How can we reconcile the Bacon who champions the advancement of the sciences and gradual human control over nature with the Bacon who speaks so contemptuously of the human intellect and preaches humility in philosophical endeavor? They are not easily reconciled, for there is conflict within Bacon's philosophy itself. Two different Bacons appear in the *Advancement* and the *Novum organum*: Bacon the empiricist, and Bacon the satirist. The former is traditional and inductivist, while the latter is more radical, satirical, and subversive. Furthermore, the empiricist Bacon stresses observation and experiment as ways to correct for subjective biases. As a religious empiricist he prays that eventually induction will "extract[] from knowledge the poison infused by the serpent which swells and inflates the human mind."[77] Bacon the satirist, by contrast, maintains that some of the mind's idols are innate, "inherent in the nature of the intellect itself, which is found to be much more prone to error than the senses."[78] Because these innate idols cannot be removed, we must "indict them, and . . . Expose and condemn the mind's insidious force."[79] Bacon the satirist recognizes that the psyche is forged in the fires of social pressures that necessarily affect the development of an institutionalized and collaborative science. In the inauguration of the Royal Society, Thomas Sprat and Abraham Cowley celebrate Bacon the empiricist. This book shines a spotlight on Bacon the satirist and traces his legacy through the novel.[80]

When scholars of eighteenth-century literature deal with the empiricist Bacon linked to the "projectors," they emphasize his belief that the New Science "promised to 'repair' man's divinely ordained 'kingdom over the creatures,' and therefore to restore his prelapsarian transcendence above and power over the natural world."[81] For human transcendence over nature to occur, philosophers had to recognize and overcome certain problems with the human mind. According to Kristin Girten:

> Though it is true, as Lorraine Daston and Peter Galison have argued, that the term "objectivity" does not accrue its current meaning as a "view from nowhere" until near the end of the nineteenth century, Bacon did envision a philosophical state of mind that accords with the kind of level and unbiased point of view generally expected of the objective scientist at least since the end of the nineteenth century. It is in his discussion of

"the idols" that Bacon elaborates why such an aloof and detached psychological state is required for the empirical method he seeks to establish.[82]

Here Girten speaks to where, in the best-case scenario, Bacon wanted the scientific mind to go.[83] But Bacon the Renaissance rhetorician also thinks there are rhetorical advantages to where the mind is. Unbiased views and "detached" psychological states would, of course, be ideal, but until we achieve them (if possible), we need to manipulate the egos we have to get minds to guide us in the directions we need to be guided.

Recognizing this conflict within Bacon's philosophy can help us account for the differences in how later writers such as John Locke and Jonathan Swift understood the connection between maxim and mind in accounts of empirical practices. Locke would repeat Bacon's worry about the misleading certainty of maximic claims within logical proofs, but in his colorful attack on innate ideas in the *Essay* he also distanced maxims from representations of mental operations. Swift, by contrast, as a satirist, would ironically follow Bacon in his willingness to align human inwardness with a maximic incoherence on the page designed to humble the reader. For Bacon, however, probative maximizing could be epistemologically helpful in confronting readers with the incompleteness and uncertainty of their knowledge. Girten and others see Bacon the empiricist attempting to detangle human from nature by cultivating a psychological state of detachment in the masculine scientist. Yet Bacon the satirist allowed the chaotic and mysterious world beyond man's control to extend to man's own interior. Aristotelian Scholastic philosophers had enclosed themselves in this inner realm of chaos, constructing figurative "walls" that segregated them even further from real, natural phenomena. Probative aphorizing was a method of wall-breaking that would first liberate thought before its redirection with the help of experiments and nature's own surprising exceptionality.

The early novels I analyze follow Bacon the satirist in that both their representations of inwardness *and* their resistance to such representations emerge from their writers' desire to eschew confining privacy by embracing a useful publicity. Brad Pasanek reminds us in *Metaphors of Mind* that people "identify their minds with the places where they conduct their secret thoughts," but "there is no special objectless, inner place in which thought happens."[84] Consider again Bacon's Scholastics laboring in the darkness of their monastic cells. Initially Bacon's comparison is as follows: Just as the cell inters the body, Aristotle's works inter the mind. A book is like a cell in which the mind can be entombed. Aristotelian thought and writing occur in an interior, walled space of isolation and oppressive privacy. The men's intellects are trapped, claustrophobic. Bacon's is only one of many possible metaphorical options for figuring the mind as a dark, private room. Pasanek, for example, points out that the mind as interior could also be a "dark closet" in the archaic sense of "a room in which a chamber pot or close stool is stored."[85] This latter connotation is not entirely absent from Bacon's passage. Although Bacon's spiders produce "cobwebs of learning," as Jonathan Swift makes clear in his remix of Bacon's analogy in *The Battel of the Books* (1704), the silk emerges from the rear end. In deciding on the value of the Spider versus the Bee, Swift's Aesop declares: "*Erect your Schemes with as much Method and Skill as you please; yet, if the materials be nothing but Dirt, spun out of your own entrails (the Guts of* Modern *Brains) the Edifice will conclude at last in a Cobweb: The Duration of which, like that of other* Spiders *Webs, may be imputed to their being forgotten, or neglected, or hid in a Corner.*"[86] Swift, of course, shifts the target of Bacon's original analogy. His Spider represents the Moderns, and his Bee the Ancients, whom he prefers.[87] Interestingly, Swift seems to leave room for including himself among the Moderns who contribute to a genuine vein of satire, born of the spider's "poison," created by "*feeding upon the* Insects *and* Vermin *of the Age.*"[88]

Swift's satire objectifies and externalizes the motions of the mind without divorcing those motions from a sense of privacy and shame. To see someone's private writing is to see them laid "open," because you are staring at the issue of their entrails, the waste that emerges from "*the Guts of* Modern *Brains.*" This Modern Spider lives in the library, and if it is walled, it is also crowded with people, a public space enabling dissension and plenty of ignorance. By following into eighteenth-century fiction Bacon's and La Rochefoucauld's deflation of private interiority and their valuation of the publicly not-known, *Maxims and the Mind* presents an eighteenth-century novel that is fascinated by minds because what is not *yet* in a mind joins what is there in order to generate worlds.

Bacon's spiders and Swift's more literal Spider enable metaphorical constructions of the mind as an interior space where knowledge is problematically produced. Here inwardness prevents the individual from engaging usefully with the ongoing progress of knowledge about the natural world. Yet, both in his fascinating metaphors of mind and in his willingness textually to manifest the unknown in the service of cultivating the known, Bacon bequeathed to eighteenth-century fictionists a rich array of possibilities for imagining the mind as a space of both productivity and inadequacy. Furthermore, Bacon's aphoristic method lays the groundwork for modes of mind-writing that were social and less about an individual than about the wider world. In scholarship on the early novel inwardness is taken primarily to be a relic of subjectivity—the "inner" truth of the self. Yet Baconian empiricism is a project of knowing the world that acknowledges our lack of mastery over our intellectual interiors. From the perspective of an individual thinker, understanding is always partial. Others must engage for the project to extend beyond the self.

Recent post-Watt history and theory of the novel in English has emphasized how the emerging genre resisted the construction of what would become known in the scholarship as the self-determining liberal subject. Macpherson, for example, demonstrates that a logic of strict liability, which held persons responsible for harms they did not intend, underwrites the "realist aesthetic" of the early novel such that these narratives cannot possibly sponsor, as they are said to do, liberal individualism.[89] "In its strictest formulation," Macpherson writes, "liability law is a way of thinking about obligation that cannot but be indifferent to the self-directed, prudential calculations suggested by the term 'interest.'"[90] The so-called liberal subject is self-interested and determined personally to evolve or improve, and yet, in early novels in English, such self-interested improvement is consistently thwarted. In *Born Yesterday: Inexperience and the Early Realist Novel,* Stephanie Insley Hershinow attends to the counterfactual narratives of early realist fiction fueled by characters (typically women) who resist the demands of personal growth and accommodation to the "real" world through experience. *Born Yesterday* critiques "the too-easy alignment of novelistic form with individual formation" by foregrounding the early novel's preoccupation with "consistently inexperienced novelistic characters."[91] Hershinow and Macpherson have thus pried apart novel and modern liberal subject. In their accounts novelistic subjects are women, display surfaces as opposed to depths, are aligned with others through accident rather than choice, and resist personal growth. In turn, the novel reader is confronted with a less rosy view of the relationship between reading, life, and liberation.

What these accounts do not explore, however, is how the attainment of ultimate mastery through temporary submission to nature promised by

Bacon is quite literally *formally* rejected in many early novels that nonetheless aim to "submit" to nature through the adoption of certain mimetic practices. The maxim is a formal refusal fully to capitulate to techniques of what many still call, following Watt, "formal realism." Maxims in prose romance had a literary-moral purpose: they were guiding hands that ensured a reader would not stray from the preordained path of narrative interpretation. In the late seventeenth-century and eighteenth-century fiction I examine, however, they do not have such a didactic function. Instead, maxims appear in the lead-up to and the wake of inexplicable violence. These maximic moments encourage readers to abandon their intellectual defenses—defenses that can lead us to impose rational explanations onto scenes of suffering at the expense of taking in the horror we are witnessing. Maxims that fail to apply to narrative events inspire readers to explore a range of issues related to the question of what type of knowledge fiction can produce and what ignorance in fiction makes possible: Can and should fiction represent real trauma? How does one represent the private thought of an individual mind, and should one represent it? Ultimately, maxims in novels mark the limits of human understanding and trouble assumptions built into novelistic projects of knowledge expansion. Maxims articulate the limits at the center rather than the periphery of the novel.

Each of my chapters examines methods of maximizing developed by one early novelist. Each chapter's close readings unfold against the background of Enlightenment philosophical debates regarding knowledge production and knowledge access. Chapter 1 examines one of the earliest English translations of La Rochefoucauld's *Maximes:* Aphra Behn's 1685 *Seneca Unmasqued*. In this chapter, the conflict between Stoicism and anti-Stoicism provides the book with another foundation for discussing inwardness as either a requirement for or an impediment to moral, political, and natural philosophical consensus. In Behn's hands, Bacon's "knowledge broken" and La Rochefoucauld's maxims retain some of their modern scientific ethos (we need methods of learning that will encourage doubt and examination) *and* contain a critique of that ethos (self-love and private interest are at the bottom of everything, not the selfless pursuit of knowledge for the benefit of the public good). When Behn philosophizes through maxims, she does so to consider the limits of our knowledge of the real, a method she opens to women as well as men, and that draws on personal experience, imagination, and desire. This method of unknowing is endless, a pursuit of a "secret knowledge" that is never satisfied. It is pleasurable because it is humbling and relinquishes opportunities for personal gain.

The book's second chapter focuses on Jonathan Swift's engagement with maximic writing, first as part of his explicit satire of Baconian philosophy in *A Tale of a Tub* (1704, 1710) and later as part of an extended consideration of the moral boundaries of the human in *Gulliver's Travels*. This chapter's analysis of maxims helps us adjust our conceptual categories to understand how *sententiae* could signify not moral clarity but energetic intellectual breakdown. Swift is allied with Bacon on one point: the dangerous restlessness of the human mind. Through readings of the *Tale* and *Gulliver's Travels*, the chapter argues that Swift's maxims create effects of inwardness in his fictions—effects caused by the maxims' operation as satires on empiricist methods. This view of the rise of protonovelistic interiority adds considerable nuance to the traditional narrative of the rise of the psychological novel.

If Swift created novelistic interiority in order to malign it, Richardson uses maxims in *Clarissa, or, The History of a Young Lady* (1747–48) to consider inaccessible interiors. This is the topic of the book's third chapter. The chapter contends that Richardson's deployment of literary generalization in *Clarissa* is central to the novel's realism, which aims to use language to reference the world its readers inhabit, a world that includes but is not limited to useable fictions. This insight demands that we revisit canonical accounts of early "provisional" realism, including assumptions regarding how realist reference arises from empirical programs and in this way contains functional explanatory power. *Clarissa*'s realism—which coalesces around Paper X and the formally similar "meditations" that Clarissa begins composing at the same time—does not assume that reality is limited to a material world that science makes speak and that necessarily propels human progress.

This chapter deepens our understanding of the paradoxes of psychological realism. As Frances Ferguson first argued in "Rape and the Rise of the Novel" (1987), the formal conventions of the psychological novel create the *illusion* that readers can secure knowledge of a (fictional) person's mind, yet such conventions are also predicated on the liberal norm of respecting individual privacy. In other words, the novel seems to make "speak" an individual who has the right to remain silent and unmoved. The chapter demonstrates the importance of this paradox to the early novel's engagement with a material reality beyond the human.

By the beginning of the nineteenth century, maxims were increasingly under fire by literary and philosophical "gentlemen" devoted to advancing the status of a subject liberated from the world by his engagement with the imaginative arts. The novels of Jane Austen, in which witty narrators appropriate the maximic speech of common (i.e., vulgar) minds are both a case in point of the maxim's waning status in the early nineteenth century and a final stand of resistance against pro-interiority ideals. An examination of women maximizers in Austen's *Northanger Abbey* (1817) and *Pride and Prejudice* (1813) constitutes the bulk of this chapter. I analyze Austen's portrayal of consciousnesses that fail to grasp—let alone master—the laws of the physical universe, and her interest in how and why, when the world disappoints us, we turn to the impressions books have made on us. The chapter concludes with a consideration of worthless knowledge as theorized and produced in *Pride and Prejudice*—Austen's particular early nineteenth-century spin on "knowledge broken."

1
"Odd Fantastick Maxims"
BEHN'S PARTIAL KNOWLEDGE OF LOVE

APHRA BEHN was keenly aware of the inaccessibility of seventeenth-century scientific debates to educated women to whom traditional philosophical schooling had largely been closed. In her work—especially that influenced by Lucretian Epicureanism—she draws on the Enlightenment-era belief that women are more sensitive and thus closer to nature, because it allows her to claim the authority of an intimate, embodied understanding of the physical and psychological mechanics of desire. Behn, in other words, is among the women author-philosophers who positioned themselves against the early modern scientific ideal of the male modest witness. As discussed in this book's introduction, historians of the connection between literature and science in the long eighteenth century tend to follow Shapin and Schaffer's work from the 1980s, relying on the concept of the modest witness and virtual witnessing (and critiques thereof) to understand the late seventeenth-century social production of scientific fact. In *Sensitive Witnesses,* one study in this tradition, Kristin Girten explicitly traces seventeenth-century masculine scientific modesty back to Francis Bacon's concern with the human mind's distortions and forward to nineteenth-century scientific objectivity. Girten argues that Behn's writing reflects her opposition to masculine scientific modesty, because Behn believes in the vulnerability of all humans to our environments, a vulnerability that can be turned into a philosophical asset, especially for women. Behn posits a distinctly *immodest,* porous relationship between subject and object and subject and subject. Such a relationship reveals an alternative version of the story of the rise of the "modern self," which typically begins with the alienation and enclosure of one's property in oneself described by Locke in the *Second Treatise of Government* (1689). Such a Lockean ideal of self-ownership was exclusive, of course, to propertied men.

Girten's theorization of what she calls "sensitive witnessing" opens our view onto a previously unrecognized literary landscape of the long eighteenth century. Here, women author-philosophers such as Behn precisely critique the operation and stakes of masculine empirical values and methods, offering feminist alternatives. This chapter will attend to Behn's feminist materialism while presenting an account of her maxims informed by Bacon's aphoristic method of presenting "knowledge broken." Because in what follows I do not foreground the role of witnessing in early empiricism, my version of the relationship between Behn's literary inventiveness and her feminist-materialist philosophy looks different than Girten's, though the two are not incompatible. I see Girten's work as importantly focused on the long-term effects of the human aloofness from nature posited by empirical objectivity. By contrast, I attend to the philosophical consequences of Behn's playfully inappropriate translation and deployment of La Rochefoucauld's maxims in the context of her feminist-materialist account of love. The result is a seventeenth-century literary experiment with maxims as a method for communicating natural knowledge as a work in progress. Using this book's account of Baconian aphorizing, we can newly appreciate this method as "probative" and thus part of an empirical project. Behn believed that love was a force operative in and central to the material reality of the natural world. Behn's "Love," structured by nonhuman laws that we can access only indirectly, operates along the lines of Bacon's "Nature." I capitalize "Love" and "Nature" throughout the remainder of this chapter to mark Behn's understanding of their autonomy from the limited human perspective. For Behn, our understanding of Love grows slowly and collaboratively, just as our understanding of Nature does. Humans yearn to become one with Love—to lose ourselves within it—or to control it, and yet Love is not something we can master or own. We must develop practices to identify and follow Love's dictates in us. Learning how to unknow the stifling conventions of the traditional "arts" of Love is such a practice, and one important to Behn's project of communicating gradual knowledge growth through maximic writing. In using the maxim form to cultivate unknowing, Behn is not so much resisting the privatized interiority of the modern self as she is offering an inwardness associated with the power of owning up to the frequent inadequacy of our written words to record our true and often contradictory feelings.

Behn was a frequent translator of popular French works, and when François de La Rochefoucauld's *Réflexions ou Sentences et Maximes morales* (*Moral Reflections or Sententiae and Maxims*), first published in

French in 1664, crossed the Channel, she produced one of the first English translations, *Seneca Unmasqued, or, Moral Reflections* (1685).¹ The title couples a Stoic philosopher with a term suggestive of derision: To "unmasque" is to reveal truths hidden under an apparently smooth, unfeeling surface. Behn published *Seneca Unmasqued* in 1685, the same year in which she released the second of three volumes of *Love-Letters Between a Nobleman and His Sister* and several years before the publication of her amatory fictions of the late 1680s, such as *The Fair Jilt* (1688) and *The History of the Nun* (1689), and of her scientific satires and translations, such as *The Emperor of the Moon* (1687) and *A Discovery of New Worlds* (1688). Her engagement with the maxim as a modern literary form thus occurs alongside her composition of amatory prose fiction and her satirical engagement with new scientific practices.

Given that amatory fiction shares conventions with older forms of prose romance, one might argue that the presence within these narratives of an apparently didactic element such as the maxim is unsurprising. Maxims encapsulate the abstract ideals with which heroines of romance tend to be associated. Yet to believe so would be to overlook the importance of the *anti*heroism of the modern, La Rochefoucauldian maxim to the type of amatory secret histories Behn practiced. This type of fiction rejects the ideals of traditional heroic romance; it foregrounds the erotic dynamic between woman writer and her passive male hero; and it flirts with the idea of surrendering to Love—to the real truth of passion—as both a rejection of existing social pressures imposed on women and an act of modern, natural philosophical inquiry. Behn's late, loose translations of French maxims and prose romance grant the authority of an openly inquisitive, unknowing relationship to Love to the woman who enacts such a surrender. Ultimately, Behn's work suggests that the fullest version of this relationship and its epistemological promise exists in its endless extension to readers.

Scholars have not typically connected Behn's amatory fiction—let alone her amatory maxims—to her interest in natural philosophy beyond identifying in the former various critiques of the social and moral organization of the latter. This chapter makes the case for the literary significance that emerged from the connection Behn drew between Love and Nature by turning to her translation of La Rochefoucauld's *Maximes*. In this translation, Behn rearranges La Rochefoucauld's original ordering of individual maxims by moving all the maxims on Love to the end of the collection so they comprise one of several concluding sections. It is in this lengthy closing section containing maxims on Love that Behn repeatedly injects (apparently indiscriminately) the first-person feminine pronoun into the impersonal maxim form—for example: La Rochefoucauld's "It is very hard to break with someone, when you no longer love each other" becomes "I have found a great deal of trouble to break off with those I lov'd no more" (*SU* 120).[2] These disruptions of the established conventions of La Rochefoucauld's collection and the impersonal form of the maxim creates the illusion for readers that they are encountering personal secrets of one of the *Maximes'* readers (namely, the translator of this English edition).[3] Behn layers this surprising unconventionality atop the collection's message regarding the deceptive appearances of virtue, inviting future readers to see the collection as potentially also about themselves and the discrepancy between their own thoughts and actions.

Importantly, Behn's maximic self-display is part of a project of natural philosophical inquiry with Love standing in for Nature. Yet, unlike the seventeenth-century modest witness, who established the credibility of his written observations by creating a disinterested persona and erecting a discursive boundary between subject and object, or self and environment, Behn uses her maximic practice to demonstrate that when we write to disrupt convention and display individual error or ignorance, we enable our reader to experience the satisfaction of recognizing that they too possess a similar unknowing that can fuel inquiry.

By inviting readers into a literary self-examination without the potential for mastery or ownership of self—an examination that is *also* an engagement with the other—Behn reimages the role of interest in the contract between the writer and reader of amatory works. As someone who composed dedication upon dedication, Behn was well versed in interested speech in the form of conventional flattery. By the late 1680s Behn, I think, was tired of flattering and giving in (tired, even, of flattering herself), and she saw a way around it that would potentially still engage readers. Unlike late seventeenth-century gentleman natural philosophers who sought to cultivate and project a modest, disinterested perspective to earn recognition as credible witnesses of experiments, Behn suggests (following La Rochefoucauld) that we cannot trust even our apparent disinterest to be uninterested.[4] Behn's position here goes beyond what she borrows from La Rochefoucauld. As a feminist materialist she rejects the version of epistemological disinterest that entails detachment of the observing subject from the object of study.[5] Furthermore, Behn sees a possibility for avoiding the lies (intended or not) that follow claims of a disinterested perspective. In *Seneca Unmasqued* she eschews a prose style that attempts the unmediated representation of observed experience or the movements of thought inside the mind. Such a style hopes to minimize the role of the writer by rendering him detached and disinterested. In eschewing this style, however, Behn does not reject altogether the goal of communicating natural knowledge. She follows Bacon in adopting a maximic probative method designed to communicate such knowledge in growth. She departs from him, however, in aligning such a method with that pursued by lovers. The lover aims not to control the beloved's thoughts and beliefs. Rather, she aims to plant the seeds of Love that grew in her own heart so that they might grow in the recipient's. There is no reason for her addressed lover or for us to believe her, because she is not asking for our belief. She is inviting us to continue the work of trying to understand the human relationship to Love. To examine and inquire into Love is to show it proper respect.

Ultimately, I account for Behn's interest in maxims as an interest in a situated method of philosophical unknowing that metabolizes individual desires for the purpose of glimpsing a truth that we may never understand, though we may feel as though we've briefly touched it. Submitting to Love, which is akin to submitting to Nature, entails accepting Love's secrets as incomplete and only inadequately expressed, at least in the form in which they can be published. Love's secrets are ultimately unknowable for any individual, although the active practice of unknowing can lead to "touching" them.

This submission to Love is ultimately enjoyed in secret by a reader who can become a writer (whether literally or figuratively) of her own partial knowledge of the passions. Behn follows both Bacon and La Rochefoucauld in exploring how maxims might retain some of their modern scientific ethos while at the same time critiquing that ethos. Writing in maxims is a probative method open to women and men, one that encourages learning through doubt and examination. Yet even probative methods cannot ensure the pure, selfless pursuit of knowledge for the benefit of the public good. Self-love and private interest are still at the bottom of everything. Behn's consideration of our knowledge of the real and how it is limited draws on personal experience, imagination, and desire. This method of unknowing is endless, a pleasurable pursuit of a desire for knowledge that relinquishes opportunities for personal gain and is never fully satisfied.

The Duke and I

Playful provocations suggestive of a battle for predominance suffuse the dedication to "Lysander" that opens Behn's *Seneca Unmasqued*.[6] Behn has quite a lot to say to Lysander about how she is writing for herself alone. "Lysander" is the pastoral pet name for either John Hoyle, who was trained in the law and with whom Behn had a relationship, or for a different man, whom she had previously dubbed "Alexis."[7] Behn signs the dedication with her own pastoral alias, "Astrea," although (as will be discussed later) a second alias emerges toward the end of the translation itself. In the dedication Behn teases Lysander with reminders of their own relationship (he is, she says, always criticizing her) while suggesting that she and La Rochefoucauld, whom she calls "the Duke," possess a private understanding based on their shared attraction to maxim writing. The maxim collection, according to Behn, offers the individual writer an effortless mode of expression ungoverned by an audience's expectations and demands. "These Maxims," she writes, were "not design'd to be made publick" by either La Rochefoucauld or herself. They—and their form— are the result of a writer unbound by "Rules and Methods," which are unnecessary in writing not "designed for publick view" (*SU* xlvi–xlvii).[8] "*Monsieur* the Duke and I," she claims, write "purely for Idleness, and our own Lazy Diversion."[9] As Irwin Primer, the editor of one scholarly edition of Behn's translation of La Rochefoucauld points out, she is being playfully ironic throughout this dedication. Behn knows that as a popular playwright with financial difficulties she belongs more to the category of those who write for "advantage," as she later puts it, than of those who write for idleness.[10] Yet irony cuts both ways, and in placing herself in the same category as "the Duke" Behn blurs the apparent stylistic divide between French Neoclassical rationalism and "Irregularity and disorder" (*SU* xlvii). Her deliberately provocative phrase "Lazy Diversion" aligns this "unstudied, and undesigned way of writing," which she is careful to note is her preferred style, with that adopted by "a Courtier" as opposed to the "tortured, and wrack'd" regularized style of "the Learned" (xlvii). Indulgence and a lack of discipline are rights she and La Rochefoucauld share. Behn insists that her maximic method, at once messy and refined, receive the same respect as his.

Behn thus positions herself as part of a privileged milieu that can eschew scholarly order, but in her translation she ultimately locates the power of this position in a radical openness to forces of Nature and, in particular, the force of Love. We see such openness, she suggests, in those who write in disconnected maxims. They are unbound by customary rules of rhetoric. Her account of the desultory maximizer is unique in the analogy she draws between the style of maxim writers and the style of lovers: "I think [writers of maxims] are not concerned in such a Case to follow Rules and Methods, it being as unnecessary where People write but to ease their minds, and just as things fall into their thoughts, as to make set Speeches in Love, and study for Eloquence when there is none in Love like that of Love it self: no, at this time we left Rule and Order to those who write for advantage" (*SU* xlvii). This analogy emerges from the passage cited above, which identifies and distinguishes among motivations for writing by attending to the relative orderliness or disorderliness of the writer's method of presenting her thoughts. According to Behn, people who write for advantage, whether for honor or money, tend to follow methodical rules of arrangement, whereas writers motivated by a desire for mental relaxation are unbound by "Rules and Methods." Mental relaxation in this case is achieved by an outward turn. This outward turn, however, occurs first toward the space of writing. "Just as things fall into" thought they fall onto the page, with no attempt at order. This type of writing unwinds the mind by loosening public reason's and rhetoric's hold. In the comparison Behn draws, those "who write but to ease their minds, and just as things fall into their thoughts" have as little need for method as lovers speaking straight from the heart, for whom there is no "Eloquence" so effective in Love as "Love it self."[11] It is a striking but strange comparison, given that communication to a lover is audience-oriented and unburdening one's thoughts is a one-sided act. The connection lies in the shared lack of intention to arrange one's words into systematic order for the sake of social convention or acceptability. Pleasurable relief is the main outcome sought by both lover and maxim-writer. The mind "eases" itself—avoids pain—and the lover permits the release of uneasiness as natural "eloquence" pours toward the beloved.[12] In both cases, arrival at an end or ultimate fulfillment may be at quite a remove. Reams of writing could be necessary to achieve mental relaxation and seduction may not be swift. As we will see in the next chapter, writers other than Behn fear that such an involuntary performance will repulse rather than attract the recipient. The expression "one cannot help oneself" in the sense of temptation is particularly apt to describe this type of writing in the name of release.

Furthermore, as in all cases of metaphor, Behn's analogy communicates not merely one idea. The first element in the comparison, the unbound method of writing to ease one's mind, takes on aspects of the second element, the natural eloquence of Love's nonmethodical expression, and vice versa. The solitary writer spontaneously and indiscriminately recording thoughts becomes the communicative lover as the lover becomes more solitary, more a writer of the self.

Behn negotiates two relationships at once in this dedication: her relationship with Lysander, and her relationship with La Rochefoucauld, the author of the work she has come to know intimately through translation. In neither case is Behn's role defined by feminine passivity or a kind of modest feminine unknowing (a demure refusal to assert her own knowledge). Rather, Behn as author refuses to accept and operate according to rules of respectability, rules that control not only social behavior but also conventions of writing. Wisely, however, she situates this refusal as an imitation of a "natural," masculine aristocratic writing practice. In the dedication she suggests that the style that appears most disorderly is actually the style least selfishly geared toward private advantage and most open to a desire that extends beyond the boundaries of the self and its interests.

The Anti-Dogma of La Rochefoucauld's *Maximes*

Behn's ingenious and ironic account of the "natural eloquence" of disordered sentences justifies her suitability for the role of translator of the apparently rationalist maxims of a male French aristocrat. In this section I want to look more closely at the philosophical details of that French aristocrat's work. Ultimately, this attention to the intellectual background against which La Rochefoucauld formulated the *Maximes* will help establish the philosophical stakes of Behn's innovation with La Rochefoucauld's maximic form, linked as it is with her eschewal of feminine respectability, her acceptance of what Augustine called our "fleshly intellect," and her insistence on a woman's close relationship to Love.

The moral-philosophical problems that inspired La Rochefoucauld's turn to antisystematic form share similarities with the problems Bacon perceived within Aristotelian natural philosophy. Both La Rochefoucauld and Bacon craft their maxims as acts of resistance to a prevailing philosophy perceived to be didactic and static; such entrenched philosophies demand blind followers rather than encouraging the active collaboration among contributors that can lead to change. Bacon's aphorisms are framed in form and content against Aristotelian Scholastic logic and natural philosophy, and La Rochefoucauld's maxims attack Stoic psychology and therapeutic methods. This attitude of critique in the work of both men is not surprising. Bacon and La Rochefoucauld shared a background of classical learning and were both embedded in an early modern masculine intellectual culture that questioned received methods of knowledge production. Furthermore, both men were committed to collaborative inquiry. Bacon suggested that delivering one's partial discoveries in maxims would enable others to extend work that was greater than any single person. La Rochefoucauld, similarly, composed his maxims alongside Jacques Esprit and Madame de Sablé as part of the latter's salon. Despite their different objects of study, both Bacon and La Rochefoucauld aim to train readers to entertain the possibility that they have been wrong and have misunderstood the basics of the natural world, including the behaviors of its human inhabitants. When Behn aligns herself with La Rochefoucauld she thus aligns herself with someone comfortable attacking tradition and received knowledge. La Rochefoucauld treats human self-report with skepticism and the human passions themselves as legitimate objects of careful scrutiny and discovery. More aesthetically motivated than Bacon, however, he also suggests to each individual reader that he can learn to be impressed, even aesthetically moved, by his strangeness to himself.

Only in midlife did François de La Rochefoucauld, an aristocrat with a military background and limited experience in the literary arts, become involved in the social and literary collective that fueled his composition of the maxims that would make him famous. An early contribution to modern social theory and to moral as opposed to natural philosophy, La Rochefoucauld's *Maximes* now sits at the intersection of disciplines: psychology, sociology, and literary studies, as well as political and economic theory.[13] La Rochefoucauld's main point throughout the *Maximes* is that apparently virtuous (selfless) actions have vicious (selfish) motives, and thus what we call virtue is actually its opposite. Virtue is unmasked as passion, and philosophical equanimity is revealed to be a sham. In a letter to Esprit's brother on the topic of both his own maxims and those composed by Esprit, La Rochefoucauld explicitly conjures this project of unmasking virtue: "As the plan of both [men's collections of maxims] was to prove that the old pagan philosophers' virtue, which they trumpeted so loudly, was built on false foundations, and that man—no matter how persuaded he may be of his own merit—has in himself only deceptive appearances of virtue, with which he dazzles other people and often deceives himself (unless faith plays any part in the matter), it seems to me . . . that there has been no gross exaggeration of the miseries and contradictions of the human heart."[14] Although we cannot be sure that any coherent philosophical or theological system guided La Rochefoucauld's construction of the maxim collection as a critique of "the old pagan philosophers' virtue," in this statement, which became one basis for lawyer Henri de La Chapelle-Bessé's prefatory defense of the first authorized edition of the *Maximes*, published in 1664, we see a critique of Stoic pride.[15] Given the "miseries and contradictions of the human heart," we must, according to La Rochefoucauld, humiliate "the absurd pride that fills [the heart], and [show] it that it needs to be supported and buttressed in all respects by Christianity."

Seventeenth-century critiques of Stoic pride were informed by Augustine's sustained attack in *The City of God* on the immorality of Stoic mental virtues. Stoics held to one primary principle: accept the things you cannot control, and focus on what you can. Human freedom and virtue consist in bringing one's attitudes and opinions in line with reason.[16] This is Stoic mental virtue. Despite their principled approach, Stoics recognized that acceptance did not come easily. As Pierre Hadot puts it, they taught that despair is a natural affective response to external conditions that cause suffering, such as within "the domain of nature" "willed by fate."[17] We must prevent such despair from controlling us and our relationship to rational virtue, which is where maxims come in, as sturdy statements that maintain one steadfast in indifference. In his study of Stoic therapeutic methods, Hadot explains that, for a Stoic, "bringing to mind" a maxim is an act in the service of human freedom from circumstance: "We give up desiring that which does not depend on us and is beyond our control, so as to attach ourselves only to what depends on us: actions which are just and in conformity with reason."[18] The Stoic doctrine of the equality of human dignity, which they extended to women as well as men, was grounded in this belief in the innate human capacity for mental strength and rationality. According to Augustine, however, Stoic insistence on mental strength was simply pride, a fleshliness of intellect (intellect polluted by the sins of the flesh).[19] La Rochefoucauld's *Maximes* picks up on this critique by eviscerating human pride and representing the human virtues in which we take pride—honesty, gratitude, liberality—as "most often, only vices in disguise."[20] We are often ignorant of the cunning of our own self-love, La Rochefoucauld argues, especially when we believe we have conquered our passions.

The anti-Stoicism that underwrites La Rochefoucauld's presentation of the passions—most basically, desire and aversion, but also self-love, hope, and fear—directs us below the surface of accepted social and moral conventions. La Rochefoucauld suggests that such conventions, including conventions of writing descended from Stoic therapeutic practices with moral purposes, are tainted by human self-interest and self-love. In her introduction to her translation of the works of the ancient Stoic Epictetus, Elizabeth Carter explained that Stoic philosophers viewed "the *reasoning Faculty*" as the "most excellent and superior Faculty," a gift the Gods granted mankind so that we might judge and make "a right Use of the Appearances of Things."[21] If the passions threaten to ensnare us, reason can set us free. La Rochefoucauld ironically attacks these ideas through maxims that subvert the "appearances" of balanced form. The poised rationality of the maxims is a false first impression, for they couple impersonal subjects ("one," "we," "the man") and abstract nominalizations ("love," "envy," "pity") with surprising syntactical reversals, qualifications, superlatives, and redefinitions. One early maxim in the fifth edition of 1678 can help us understand the book's basic principles of content and form: "Passion often turns the cleverest man into a fool, and often makes the worst fools clever."[22] This sentence conveys the basic idea that we do not rule our passions through reason; our passions rule us. It goes further, however, using a chiastic structure to blur the line between foolishness and cleverness. Passions enforce rule unpredictably and irrationally, rewarding the undeserving. Redefinition often functions this way in La Rochefoucauld's maxims, inspiring a series of questions about the relationship between our words and the reality to which they supposedly refer: What, for example, can cleverness or foolishness possibly mean within a sentence that renders them indistinguishable? La Rochefoucauld's collection is also antisystematic in that the content of one individual maxim might contradict another. When coupled, they raise questions rather than generating answers. Very shortly after the aforementioned maxim, for example, we read: "Passions are the only orators who always succeed in persuading. They are, so to speak, a natural art, with infallible rules; and the most artless man who is passionate, is more persuasive than the most eloquent man who is not."[23] This passionate, artless man has a greater power of persuasion than the most illustrious orator. Is he then clever, or a fool? Because of the lack of analysis and explanation connecting the maxims, it is up to the reader to decide. The surprising reversals, designed, in part, to suggest the constant possibility of misperception, deception, and blindness, suggest our lack of knowledge of our motivations and our selves. If Stoic practices of attention—one of which involved meditating on maxims—strengthen virtue by cultivating a dispassionate detachment from what we cannot control, then this new, modern anti-Stoic therapy encourages confrontation with the realm of cause and effect of which we are a part and that does not operate solely according to our desires. La Rochefoucauld's maxims act as delightfully humbling windows onto the deceptions motivating our actions and attempts to solidify public regard. And yet the "truth" of self-deception cannot ultimately be accessed. We cannot with absolute awareness experience our own self-deceptions head-on. Instead, we can only glimpse them. We are not masters of the universe, and we can fix neither the disorder and despair nor the beauty and joy we experience and create. We can, however, surrender to this truth just beyond our understanding and just out of reach of our expression.

La Rochefoucauld's maxims are not, of course, frequently considered alongside the genre of the early novel, whether in England or France. Yet La Rochefoucauld himself made a connection between maxim and prose fiction. He had a close literary relationship and friendship with Madame de Lafayette, author of *La Princesse de Clèves* (1678), and he collaborated with her and with Jean Regnault de Segrais on the novel *Zayde* (1669–71).[24] Lafayette belongs to the tradition of amatory fiction to which Behn herself contributed, and later in this chapter I will contrast Lafayette's strategy for conjuring interiority with Behn's own, different method for exposing characters' (and readers') "secrets" through maxims. For now let me say that, while Behn translates actual maxims on Love that exist in La Rochefoucauld's original, Lafayette's novel reminds us of the philosophical and literary seriousness with which Love is treated in novels of the period.

In describing La Rochefoucauld's *Maximes* I have emphasized the philosophical (anti-Stoic) foundations of its form, which is characterized by the playful thwarting of moral and discursive conventions. I have even suggested that the maxims aim at a producing a particular effect for readers: the fleeting feeling that they have just glanced at the truth of their own self-deceptions. Viewed in this way, the *Maximes* takes on some striking similarities to Lafayette's *La Princesse de Clèves*—often called one of the first European psychological novels. *La Princesse de Clèves* is not only a historical novella the events of which are set in and around the court of Henri II in the final years of his reign (1558–59); it is also a "secret history," which one modern editor defines as a "quasi-historical [narrative] which [purports] to describe" off-the-record aspects of historical events.[25] One of the novel's earliest critics, Jean-Baptiste de Valincour, was both troubled and intrigued by the novel's seemingly ostentatious departure from conventions of historical fiction. As Nicholas Paige explains in *Before Fiction,* according to the Aristotelian rules of poetic invention faithfully followed by Lafayette's contemporaries, authorial invention within historical fiction was to be confined exclusively "within the blanks of . . . historical record." These were the blanks filled by writers with the "entanglements" of Love in order "to explain or motivate noteworthy past events, thus producing the sensation of reading . . . 'secret history.'"[26] Lafayette, however, departs from such conventions in *La Princesse,* because she seeks to compose a novel that is not beholden to the logic of gossip. *La Princesse,* Paige writes, was not a work of gossip but "a novel *about* gossip."[27] Rather than keying her heroine to an actual real-world referent, such that readers need only look to the details of the historical record to find the real origins of the sexual secrets the story narrates, Lafayette creates an "impossible princess"—not just implausible in her motivations and actions, but "factually impossible."[28] This novel creation completely "disables normal reading protocol," enabling effects such as a new "emotional bond" between reader and character.[29] I will return to this new emotional bond between reader and character below, when I discuss it in relation to Behn's translation. What is important to note here is that La Rochefoucauld and Lafayette, perhaps working alongside one another, both experiment with breaking typical genre conventions—the former of maxims, the latter of secret history—in order to disrupt the relationship between writer and reader, or, as Bacon put it much earlier in the century, between deliverer and receiver. Neither La Rochefoucauld nor Lafayette wanted superficially to satisfy their readers. They wanted them to reflect on the moral and epistemological uses to which previous genre conventions had been put. I turn again now to Behn, who has her own unique investment in disrupting the moral and epistemological uses of maximic form.

The More I Love the Readier I Am to Hate

By directing attention to the form rather than the content of the *Maximes,* the prefatory analogy that frames *Seneca Unmasqued*—that idly writing to ease one's mind is like spontaneously expressing one's Love—prepares a reader for Behn's most striking changes to the original: her rearrangement of the maxims, including a new section entitled "*Of LOVE*" (a heading that does not appear in the *Maximes*) placed near the end of her translation, and her decision to shift the rhetorical situation of some of La Rochefoucauld's utterances from authoritative address to apparently personal disclosure in this section.[30] These decisions are especially striking given the gender Behn assigns to the voice making this personal disclosure. Originally readers of the published *Maximes* assigned its authorship to a single man. On the title page within the *Miscellany* in which Behn's *Seneca Unmasqued* originally appears, authorship is assigned, by contrast, to a single woman: "*Seneca Unmasqued, or, Moral Reflections. From the French: By Mrs. A.B.*" Furthermore, in the grand tradition of masculine classical literary accomplishment, La Rochefoucauld's *Maximes* addresses a collective male audience and frequently signals the privileged exclusivity of this address by attacking one primary group excluded from it: women. Thus, La Rochefoucauld writes, "Sometimes it is less unfortunate to be deceived by your beloved than to be disillusioned by *her*,"[31] and "The more you love your beloved, the closer you are to hating *her*."[32] In her translation, Behn takes multiple, ingenious approaches to these attacks on the primary type of woman that appears in the *Maximes*: one sexually involved with a man.

While La Rochefoucauld places a number of his maxims on Love toward the beginning of his edition, Behn groups them together at the end of her translation. More notably, in this new section "*Of LOVE*" she shifts several of the pronouns of the originals from the impersonal "on" in the French (translated to an impersonal "you" in the Blackmore and Giguère translation) to a self-assertive "I" and delivers them in the voice of a feminine alias, reversing the implied gender of the speaker in the original. Behn translates one of La Rochefoucauld's most famous maxims as: "I should never have been in Love if People had not talked of Love to me" (*SU* 114). La Rochefoucauld's "The more you love your beloved, the closer you are to hating her" becomes "The more I love *Lysander*, the readier I am to hate him" (112). By introducing the proper name "Lysander," Behn lends a social position to the subject behind the first-person pronoun, elsewhere identified in these translated maxims on love as "Amynta." She additionally shifts the address of the original from a general recipient to a specific one. La Rochefoucauld's "If you think you love your beloved for her own sake, you are very much deceived" becomes, in Behn's version,

"If you believe you love *Amynta*, for the love of *Amynta*, you are much deceived" (124).³³ In separating and compiling together the maxims on Love (and its synonyms for Behn, "Passion" and "Passions"), Behn posits a difference between these generalizations and the majority of maxims in the collection and rewards with a surprise the reader that skips to the end to indulge in (the category of) Love.

I have emphasized the maxims on Love, because those are the ones in which Amynta asserts herself. Along with instances of the first-person plural "we"; the general "one"; "Man" as contrasted with "Woman,"; and abstract nouns (e.g., Virtue, Vice, Courage) as subjects, Behn maintains the use of general "man" ("l'homme") throughout the majority of her translation, even at times including the masculine pronoun where it was omitted in the original. For example, whereas Blackmore and Giguère translate La Rochefoucauld's "On n'aurait guère de plaisir si on ne se flattait jamais" as "We would have few pleasures if we never flattered ourselves," Behn renders it as "He loses much satisfaction, who does not both flatter himself, and is not flattered by others" (*SU* 30).³⁴ The impersonal "we," "you," or "one" to which La Rochefoucauld addresses his maxims suggests that each maxim's truth might be applicable to any individual. As established, however, the general individual posited by La Rochefoucauld's impersonal address is male. By the end of Behn's translation, in contrast, the reader both sees women included in this address and becomes a voyeuristic bystander to an intimate conversation that Amynta maintains with her lover throughout the translation.

Previous scholarly accounts of Behn's translation have essentialized the playfulness with which she approaches contemporary associations between literary form and gender. Irwin Primer, the editor of one of two scholarly editions of *Seneca Unmasqued*, suggests that when Behn particularizes the abstract (a decision he implicitly genders feminine) she "eroticizes" La Rochefoucauld's work, introducing a "subtext" "foreign to the duke's intentions."³⁵ "Some would call it vulgarization," Primer writes. "If she meant to shock, the shock value has long since disappeared and we are left with a mildly bawdy translation which she probably intended as wit."³⁶ Primer's condescending evaluation might lead one to believe that Behn departs further from the original than she actually does. Each of La Rochefoucauld's maxims advances a claim that is meant to capture something both true and general, the latter criteria meaning that the truth applies to numerous individual cases. This quality of his claims means that charges of untruth and unreality cannot be laid at the door of peculiar individual feelings or biases belonging to him (although his critics certainly did lay this charge at his door). Behn picks up on this dimension of La Rochefoucauld's maxims and exploits it when she selects individual maxims and translates them not only from French to English but from a general claim to a specific claim about Amynta's and Lysander's passion. Roland Barthes would agree with the interpretive legitimacy of Behn's practice, for he says of La Rochefoucauld's maxims that they can be read in two ways. When read collectively they tell the story of the author, "a kind of obsessive monologue." When read individually, however, they become truths applicable to each individual reader: "This particular maxim traverses three centuries to land on target, to tell about *me!*"[37]

The prefatory analogy speaks to a unique erotic relationship between Astrea (a.k.a. Behn) and Lysander, while the section "Of LOVE" alludes to a separate erotic relationship between Lysander and Amynta. Each relationship is defined by writing, and we readers see only half of the words exchanged. The traditional "art" of Love this is not. Rather than merely inspiring readers to reflect on and be humbled by the lack of correspondence between our moral terms (honor, courage, cleverness, foolishness) and the reality of our selfish motives, as La Rochefoucauld's maxims do, Behn's maxims on Love make these reflections dialogic, turning them into the materials of an insinuating, incomplete conversation between lover and beloved. One of La Rochefoucauld's original maxims, for example, measures the happiness of a lover in relation to the illusions he holds about his mistress. In her version, Behn spotlights masculine deceit and embeds her speaker in the deceit described. It is an awkward situational posturing. She translates "Sometimes it is less unfortunate to be deceived by your beloved than to be disillusioned by her" to "I am more happy in being deceived by *Lysander* than in being undeceived" (SU 126). The sentence, now in the present tense, is no longer hypothetical. If Amynta *knows* Lysander deceives her, then she is already undeceived. The trick is that this is no common maxim; it is directed at the deceiver himself, with the added benefit of plausible deniability (this is, after all, a translation). On yet another level, the deceiver and recipient is not only Lysander but any reader of *Seneca Unmasqued,* pulled into an intimate and provoking second-person address by other maxims in the collection. Behn's maxims on Love are not instruments merely for humbling a person's intellectual pride; they animate the passions, serving an important communicative purpose, however odd and fantastic that purpose might seem.

Behn's and La Rochefoucauld's techniques are not as far apart as Primer suggests and as we may have assumed. La Rochefoucauld devised, as Bacon did, a "broken," antisystematic form in order to force readers to break not only with a dominant philosophical system, but with the "contract of error" by which that system had been communicated across centuries, a contract in which, as readers, they had been complicit. Behn does something similar, but with La Rochefoucauld's *Maximes*. In personalizing some of the impersonal maxims, she breaks the new conventions of maxim writing he had established. Her goal was to suggest that, perhaps, we do not fully understand how such a form makes meaning for individual readers, especially for women readers. She sets out to attract and invite all such readers to the study of the literary, maximic meaning making in which she is embedded.

Writing to One's Disadvantage; or, Submitting to the Real

Having just made the case that there is an interactive dimension to Behn's maxims of Love, it is worth explaining how such a goal might undermine (or not) the *Maximes*' message regarding the deceptive appearances of virtue and the selfishness of the passions. With the possible exception of literary studies, the disciplines with which the *Maximes* was historically associated remain concerned with accounting for human behavior, particularly behavior that is taken to be "interested"—in other words, in the service of "one's own profit or advantage."[38] La Rochefoucauld writes, for example: "Self-interest speaks all kinds of languages and plays all kinds of parts—even that of disinterestedness."[39] The modern translator of La Rochefoucauld tends to render his "intérêt" as "self-interest." Yet, for La Rochefoucauld, this is interest not "in the material sense, but most often an interest in honour or glory."[40] In other words, "interest" in the *Maximes* points to the concern for reputation that was central to a seventeenth-century aristocratic ethos. A related and well-known term throughout the *Maximes* is "amour-propre" (self-love). Consider the following example: "There is no passion so powerfully ruled by self-love as love; and we are always more willing to sacrifice the peace of our beloved, than to lose our own."[41] The seventeenth-century intellectual world filtered the concepts of self-interest and self-love through ancient philosophy, theology, and early modern social and political theory. Self-interest *was* simply what was advantageous to the self, and as an ideology it elevated self above all. To borrow Augustine's framing, self-interest is the fleshliest part of our fleshly intellect.

Behn's ironic dedicatory performance of a resistance to writerly "advantage," mentioned earlier, thus gains additional complexity, because the message of the *Maximes* is that each of our acts most likely has a selfish motive. Behn insists that the author of the "original" she is translating—whom she calls "the Duke of Rushfaucave"—did not write "for trade," and neither does she (*SU* xlvi). Yet because Behn's translation tilts the original more firmly into the personal and the private (she situates it, after all, within the confines of a mutually mortifying love affair), a realm especially associated with self-interest, even this justification for her disavowal of advantage seems suspect.

If Behn, like La Rochefoucauld, aimed at humbling a reader, she sets herself up for a challenge in the literary marketplace.[42] Ros Ballaster, in her work on seventeenth- and eighteenth-century women's amatory fiction, situates Behn's amatory writing within a late seventeenth-century literary marketplace that already reflects an emerging separation of disciplines, complicated by the rise of new aesthetic paradigms. The value of the literary commodities Behn produced was determined not by the quantity or quality of truth offered but by their appeal to readers' desires. In other words, to serve herself, Behn had to cater to her readers' interests. To make this point, Ballaster draws on a theory of modern narrative seduction developed by comparativist Ross Chambers in his work on nineteenth- and early twentieth-century narrative. Chambers writes, "When narrative ceases to be (perceived as) a mode of direct communication of some pre-existing knowledge and comes instead to figure as an oblique way of raising awkward, not to say unanswerable questions, it becomes necessary for it to trade in the manipulation of desire."[43] For Chambers, narrative can either communicate "pre-existing knowledge" directly or it can indirectly ("obliquely") communicate partial knowledge ("raising awkward, not to say unanswerable questions"), but if it opts for the latter it must offer readers some new pleasure for their pains. I find this distinction a helpful point from which to approach Behn's "manipulation" of the reader's desire. Her maxims on Love suggest that she intuited that readers might find some interpretive pains pleasurable, especially in cases where "pre-existing knowledge" was made strange through indirect communication.

As we have seen, one of Bacon's critiques of existing scholarship prior to the empirical turn was that its communication of knowledge was tainted by an affective "contract of error." Probative methods of communication could ameliorate this error, productively channeling desire in another direction. Such new probative methods would deliver partial knowledge while fueling a desire to know more, and from the ground up. This method of communication empowered the reader with the right to investigate, but at the expense of the author's pride. Under nonprobative methods, Bacon writes, "glory mak[es] the author not . . . lay open his weakness, and sloth mak[es] the disciple not . . . know his strength."[44] The *Maximes*' anti-Stoic attack on human "glory," or pride, calls to mind Bacon's critique of method and suggests its relevance to *Seneca Unmasqued*. Indeed, Behn draws her addressee—both Lysander and her reader—into a conversation in which the speaker lays open her own "weakness." It is an understatement to say that Amynta delivers knowledge under nonideal conditions, for she states outright that she has been and continues to be deceived. If we apply Bacon's probative framework here, Lysander would seem to be the "disciple" coming to know his strength through his teacher's weakness, and yet he too is deceived in his love for Amynta. The relationship Behn constructs between Lysander and Amynta thus seems particularly unprofitable for either party—not just humbling, but humiliating—and yet they persist, chasing pleasure. Considered in the context of Behn's Lucretian Epicureanism and La Rochefoucauld's insistence that we are often blind to the movements of Nature outside our self-interested orbit, it seems likely that *Seneca Unmasqued* aims at a larger goal of increasing a reader's understanding beyond awakening them to the epistemologically closed circle of lover and beloved, deceiver and deceived. If Behn's maxims on Love constitute a probative method of knowledge transmission, they are probative in a sense that, if implied by Bacon's account, is not made explicit.

Behn understood philosophical knowing and unknowing to be dependent on social and cultural hierarchies for their meaning and value. As she makes clear in "To the Unknown Daphnis," her poem in praise of Thomas Creech's translation of Lucretius's *De rerum natura*, men and male-run educational institutions denied women access to knowledge by restricting their ability to learn the classical languages in which even modern forms of study proceeded. Behn writes: "Till now I curst my Sex and Education / And more the scanted customs of the Nation, / Permitting not the Female

Sex to tread / the Mighty Paths of Learned Heroes Dead" (lines 25–28).[45] At the same time, as her engagement with materialism—particularly Lucretian atomism, as Alvin Snider has outlined—demonstrates, Behn viewed Love, in the sense of sexual attraction and reproduction, as the generative force of the universe.[46] Sexual desire, for Behn, is Nature within but separate from us. We might deceive ourselves regarding another's love of us, but sexual desire can never deceive, because it only moves in one direction.

All of this led Behn to believe that disinterested philosophical inquiry was a fantasy. Behn's work beyond *Seneca Unmasqued* suggests that she rejected the idea that natural philosophers could actually be impartial, pursuing public good as separate from their own desires. Yet her interest in humiliation as epistemologically productive also brings *Seneca Unmasqued* into alignment to some extent with disinterest as understood within late seventeenth-century experimental philosophy. In its most idealized form, scientific endeavor is intended to function independently of human temporal authority and the scientist's desire for recognition or honor. Experimental philosophers had to be appropriately trained in techniques of (neutral) observation. Their minds were to become suitably "modest," a moral condition of mental purity that would extend to the written records of their experimental observations.[47] Tita Chico, who examines the literariness of the modest witness, explains that, for Shapin and Schaffer, establishing such modesty in observers was part of the Royal Society project of producing a believable account through establishing the credibility of the witnesses to experiments.[48] For Chico, a modest observer adheres to a moral value of abandoning the distractions of interest: "He embodies, paradoxically, a privileged absence, a model of spectatorship that conveys authority through its claim to be free from the limitations of self-interest."[49] When in her own study of early scientific *immodest* witnesses Kristin Girten refers to Chico's depiction of the modest witness as "a masculine figure of authority, gentility, and privilege, admired for his morality and knowledgeability and, just as notable, distinguished from women and laboring men," she identifies the modesty as "Baconian." Yet submission rather than disinterestedness is the primary brand of humility preached by Bacon to his male experimentalists. When Bacon utilizes the language of submission, he does so in the context of vying for domination. Take, for example, the third aphorism of the *Novum organum:* "Human knowledge and human power come to the same thing, because ignorance of cause frustrates effect. For Nature is conquered only by obedience; and that which in thought is a cause, is like a rule in practice."[50] We see here Bacon's signature emphasis on operative knowledge: Nature's laws teach us how to make things—how to achieve "effects." By practicing such making we chase Nature's causes, and as we catch them, we catch her. Knowledge is power for Bacon, because knowledge brings the capacity to make, to shape, reshape, and "conquer" our environment. The submission is temporary, a stepping stone toward her ultimate submission, but it is submission nonetheless.

Behn's humbling of lovers' knowledge in *Seneca Unmasqued* calls not for disinterestedness but for submission to Love, whose "interests" are foreign to our own. Love operates along the lines of Bacon's Nature, according to nonhuman laws. While Nature's disinterestedness regarding human affairs was taken as beneficial by the Royal Society—that is, if Nature's laws are different from human laws, then there may be power we can still harness by learning from her—for Behn, Love's nonhuman laws are freeing for humans, and particularly for women. The goal of inquirers into Love, Behn suggests, should be to grow their knowledge of Love as much as possible by coming to acknowledge how their particular interests weaken their comprehension of that natural force. Bacon's final domination of Nature is for human benefit, whereas Behn does not imagine that any final human domination of Love is possible.[51] There is a desire on Behn's behalf for a process that does not require public forms, or that requires new public forms. What Behn wants to grow in her reader is the practice of unknowing the self that submitting to Love makes possible.

Maxims and Amorous Fiction: The Mirror and the Lover

Behn's use of the maxim does not, in fact, radically depart from scholarly accounts of the modern aphoristic form's function. Joshua Dienstag writes, for example, that "as a result [of the aphoristic collection's disjunction], rather than emphasizing community and identity, as a dialogue does, aphoristic wisdom tends to separate its reader from his or her self and from the group of which he or she is a part."[52] If reading philosophy—particularly reading philosophy in the form of maxims—puts readers in direct physical relation with a "real," nonhuman force greater than us all while separating us from the group of which we are usually a part, then this separation could be a powerful thing for a woman writer in the late seventeenth century in England.

In her translation Behn is writing philosophy, and she philosophizes and instructs not only men, but women.[53] When Behn writes herself into the Duke's maxims, first in the dedication and later via the persona of Amynta, she particularizes what had been general. As Sarah Raff explains, epigrammaticism bore "aristocratic prestige" in France, a distinction that had influence as well in England.[54] Particularity was, by contrast, "associated not just with boorishness and sexuality but also, by the same token, with femininity."[55] "Like feminine modesty," Raff writes, "feminine generalizing was attractive for its transvestitism: it denied a woman's connection to the unreason and lawless sexuality that were supposed to be her portion, and it allayed fears of feminine otherness. Evoking both masculinity and femininity, it shared the seductive ambiguity of that portent of innocence and (sexual) knowledge, the blush."[56] Behn, however, is not just a woman generalizing, but a woman who first generalizes only to reparticularize the general.

In breaking the conventions of La Rochefoucauld's staunchly masculine and impersonal *Maximes,* Behn poses the question of how to make sense of maxims that particularize individual weaknesses in the form of secrets disclosed. If we return now to Behn's section "*Of LOVE*" from her translation of the *Maximes,* couched as it is as an anti-Stoic document, we might consider it alongside an instance of fiction for which secrets of Love take center stage. When a reader first opens Lafayette's 1678 *La Princesse de Clèves,* they are promised a tale of court intrigue during "the last years of the reign of Henri II" of France, roughly 1558–59.[57] Women are, at this time, at the center of politics. Love and power coincide, and "rivalry" and "envy" abound.[58]

As mentioned previously, Nicholas Paige sees *La Princesse de Clèves* as an outlier within the history of the development of modern novelistic fiction. Lafayette's primary intention in molding the Princess was not to create a "deep" character in the modern sense. Rather, Paige argues, she avoided conventions of Aristotelian poetic invention in order to create a tale about court gossip and sexual intrigue that would not participate in the very structure of that intrigue itself.[59] The absurdity of the princess's violations of convention (e.g., by revealing to her husband that she loves another and then revealing that love to the man that is not her husband) breaks the spell of critical distance between character and reader that is maintained within ethical modes of interpreting fictional character. Neither opportunities for judging the appropriateness of characters' actions nor the fame of real historical actors could be "the . . . basis of readerly interest."[60] What Lafayette's early critic, Valincour, noticed was that a new foundation of readerly interest arose in their place. The Princess does not abide by any coherent philosophy (e.g., of conduct in Love), which makes it almost impossible for readers to sustain a careful analysis and evaluation of her social tactics. She displays instead a level of "social incompetence" never before seen in the historical novella.[61] Surprisingly, the result is an increasing identification of the reader with the heroine's experience. As Paige puts it, summarizing Valincour's own reaction to the novel, "Lafayette's achievement is immediately described as the *expression* of things that everyone has already felt."[62] What are these incredibly common feelings and experiences, according to Valincour? Not knowing what to say or do in a social situation, and being almost paralyzed by it. Such moments inspire fear, self-doubt, and a desire to hide both from others and ourselves. The Princess is defined by "her inability to act in her best interests"—by an "inadequation between interior and exterior, feelings and actions."[63] This is what strikes readers as most like them. Lafayette's representation of this inadequation drives the engine of readerly identification. The subject of the story is not the Princess, but the reader's own self. It is only because of this novel source of readerly identification that we can and should call *La Princesse de Clèves* a psychological novel.

I see more than a little structural similarity between the Princess's inadequation of interior and exterior and that of Behn as a private *person* and as a public *writer*. In the dedication writing under her alias Astrea, Behn writes as if at ease, perfectly comfortable with the "natural" eloquence of Love. Yet, in the "*Of LOVE*" section of the translation, waves show up on the treatise's calm exterior as the personalized maxims emerge, ostensibly belonging to another persona, Amynta. The gentility and authority of an impersonal maxim such as La Rochefoucauld's "If you think you love your beloved for her own sake, you are very much deceived" becomes bungled as social performance, apparently revealing too much: "If you believe you love *Amynta*, for the love of *Amynta*, you are much deceived" (*SU* 124).[64] Behn puts pressure on the possibility that, in the case of Love, the writer might *not* write in her own best interest. Implied is a disruption of mental equanimity, and the maxims register that blip as a surface flaw.

Behn's technique can be seen, however, as one designed to draw in readers. Readers themselves can become writers, adapting, particularizing, and reordering the maxims as they see fit. The invitation to readers is to act by writing something they almost know, at least halfway: that they themselves can be inadequate, can fail to adhere to appropriate social conventions. In such failure, however, is proximity to Nature and to Love, in all its natural art. This recognition of oneself in the mirror of the text is comforting, not upsetting or disappointing. It is as if the self's inadequacy, in this textual instantiation for all to see, is for the reader privately felt to be adequate, perhaps for the first time.

Someone who reads *Seneca Unmasqued* in this way, as a warped and probative secret history into each reader's own disjunction between amatory feelings and public actions gains an epistemological intimacy with Love that leads not to scientific knowledge proper as defined by Bacon or Locke—as operative or probable, respectively—but rather to a tenuous contact with the mysterious secrets of the unknown. We see Behn build on this idea in her 1686 translation of Balthazar de Bonnecorse's prose fiction *La Montre,* which Behn titles *The Lover's Watch: or, The Art of Making Love.*[65] In her translation of Bonnecorse's text, Behn pairs one of La Rochefoucauld's maxims on Love with an insistence on Love's reliance on secrecy for its integrity.[66] *La Montre d'Amour* is essentially epistolary in format. Bonnecorse presents the work as a collection of correspondence between two lovers, Iris and Damon. It begins with a bit of background from an editor: Damon has demanded "a *Discretion*" from his lover Iris, who is away "for some Months" in the country. Iris responds with a "Watch," a long combination prose and verse epistle containing the rules Damon must follow each hour of every day that he is apart from her. Eventually Damon responds with a "case" for the "watch," followed by a "looking-glass" in which Iris can view herself, and only herself. Iris writes to her lover Damon that during the hour of four o'clock in the afternoon Damon must endure visits and "general conversation." During this time he must not, she insists, disclose to others his love for her. The final two lines of her poem in this section reads: "*Love can his Joys no longer call his own, / Than the dear Secret's kept unknown.*"[67] Iris goes on to elaborate on the relationship between Love and secrecy, the way in which Love's joys depend on limited access to the knowledge of another's feelings. She writes: "There is nothing more true than those two last Lines: and that Love ceases to be a Pleasure, when it ceases to be a Secret, and one you ought to keep sacred: For the World, who never makes a right Judgment of things, will misinterpret Love, as they do Religion; every one judging it, according to the Notion he has of it, or the Talent of his Sense. *Love* (as a great Duke said) *is like Apparitions; every one talks of them, but few have seen 'em.*"[68] By Iris's logic, Love's pleasure will dissolve under the world's gaze. This is because a misinterpretation or misjudgment of the authenticity of the feeling of Love or the object of one's Love will destroy the power, authority, and autonomy of the experience. The only correct judges—the only persons with access to the thing itself—are the lovers. This is not, however, because Love exists only in their minds. Here things get interesting, for Behn turns to "a great Duke": the Duc de La Rochefoucauld, whose *Maximes* she had recently translated, to explain the object of her character's study: Love. Behn produces almost exactly the distinctive English version of this maxim (number 76 in La Rochefoucauld's fourth edition of 1675, which Behn used for her translation) that she included in *Seneca Unmasqued,* where it appears as "It is with Love as with Apparitions, all the World talks of 'em, but few have seen 'em."[69] Both Love in its pure sense and true lovers are fewer than we think. As for Love, "there is nothing so nice, or difficult, to be rightly comprehended." Even worse, perhaps, each person thinks she knows Love, but most of us are mistaken, for "He [Love] can fit himself to all Hearts, being the greatest Flatterer in the World: And he possesses every one with a Confidence, that they are in the number of his Elect; and they think they know him perfectly, when nothing but the Spirits refined possess him in his Excellency."[70] Here Behn comes to the origin of maxims: "From this difference of Love, in different Souls, proceed those odd fantastick Maxims, which so many hold of so different kinds: And this makes the most innocent Pleasures pass oftentimes for Crimes, with the unjudging Croud, who call themselves Lovers."[71] For Behn, Love is a thing sacred, requiring secrecy, a refusal to let outsiders pry. It is as if the public gaze threatens Love's existence. But Behn's own choice to place this insistence on privacy on a printed page designed for a reader's private consumption gestures at her interest in expanding access to the unknown—in initiating readers into its embrace.[72]

There is an important difference between the role inwardness plays in motivating Francis Bacon's early seventeenth-century account of aphoristic writing as a probative method and in Behn's claim that maxims are the natural forms that flow from the pen of a someone writing to "ease" her mind. Neither are attempts to imitate inwardness through form. Rather, they are attempts to use form to bypass inwardness's effects on writing, effects that could distort communication and the gradual knowledge growth that relied on it (knowledge of Nature in Bacon's case and knowledge of Love in Behn's). Bacon, of course, as discussed in the introduction, was skeptical of claims that the human mind could naturally produce anything useful when relying solely on its own resources. Aphorisms were a response to the individual mind's weaknesses. They would disrupt old patterns of thought with provocative gaps and guide the reader's thinking toward the natural world. For La Rochefoucauld and Behn, our inner motivations are many, contradictory, and obscured. Maxims express this indirectly while turning us outward, toward an audience other than the self. What is inside the mind does not match its external expressions, but (and this is Behn's contribution) writing without method—by maxim—can, paradoxically, ease this inadequacy by expressing it.

Scholars have argued that in Behn's amatory fiction, the individual is consumed in the fiery passion of Love. Yet we find individuation in Behn's reflections on Love precisely where we would not look for it: in her translations of another writer and her insertion of particularity in maxims formally marked by generality. This mode of individuation involves no mind containing secrets a novelist can plumb. Individuation occurs in the authorizing voice willing to undermine itself. It is an individuation open and capable of possession by anyone. The abstract "catching," "seeing" aslant, or conceiving of our own blindness through reading maxims into our experience is an individualizing task.

Scholars tend to use the language of servility and abjection to describe Behn's attitude toward lovers and the annihilation of self at which she supposedly aims. Susan Staves, for example, writes:

> More clearly than in the poetry, one sees in fictions like *The Fair Jilt* that for Behn love entails conquest and domination. Up to a point, one can share the excitement of feminist critics who celebrate Behn's gender-bending reversals as giving agency to female characters and as destabilizing the received orthodoxies of gender. To my mind, though, there is a crazed absolutism in Behn's exaltation of a supposedly noble love, like Tarquin's, that will baulk at nothing to satisfy the demands of a criminal beloved. The duty and desire of a lover in romance to serve a virtuous beloved have become dissociated from the ideals instantiated by love objects in romance, surviving as pure abjection.[73]

I would argue, however, that humble submission is an attitude Behn and her characters perform in relation to Nature-like Love itself. Bacon's Nature is similar to Behn's Love in that both of these writers use these respective abstractions to rethink the dynamic of domination and submission typically conceived of as a social relation among people. Behn was aware that culturally accepted displays of privacy, such as immersive reading, had become associated with feminine secrecy.[74] For femininity to be associated with knowledge, especially the experiential knowledge so valuable to the new scientists, absolute secrecy ought to be avoided, and a partial openness—even an indirect one—embraced.

2
The Maxims of Swift's Psychological Fiction

NEAR THE END of his narrative, Gulliver assures us of his veracity: "I imposed on myself as a Maxim, never to be swerved from, that I would *strictly adhere to Truth.*"[1] He then delivers a quotation from the *Aeneid* to further his claim:

> —Nec si miserum Fortuna Sinonem
> Finxit, vanum etiam, mendacemque improba finget
>
> (Nor, if Fortune has made Sinon miserable, will she also in her spite make him false and lying).[2]

Unfortunately, Gulliver has forgotten that Sinon is a liar—the very Greek who says "the Thing which was not" to get the wooden horse into Troy. Quotation does not reinforce maxim but contradicts it. This absurd arrangement of authorized instruction illuminates the error that is the object of Swift's satire at this point in *Gulliver's Travels* (1726): Gulliver's struggle to wrest meaning from experience on the strength of his own intelligence. The ironic combination of maxim and quotation mocks this vain attempt by suggesting an incongruity between what Gulliver writes and what he understands. Gulliver's mind is moving, but going nowhere.

Rather than expressing collective, public, even self-evidently rational ideas, Gulliver's maxims expose a lack of insight, an aspect of his private being. Gulliver is a naive empiricist attempting to access truth by recording his own experience, and yet, when he writes, he ironically reveals his intellectual disorder.[3] Gulliver's thoughts accumulate but do not add up, creating the sense that an active failure of thought is unfolding as we read.

Scholars have examined Swift's satirical attacks on contemporaries who utilized modern scientific procedures. When it comes to the late seventeenth- and early eighteenth-century taste for empirical precision,

Swift was a skeptic. He mocked writers who applied new empirical and organizational methods—such as description, collection, and categorization—to the study of history and literature. Such methods, Swift believed, bypassed thought, making room for essentially noncognitive practices.

This chapter pays close attention to Swift's satire of one specific method for conveying partial, empirical knowledge: the aphorizing of Francis Bacon. Swift's imagination was drawn to Bacon and, in particular, to Baconian aphorizing. We might easily categorize Bacon as one of the conqueror-innovators mentioned in *A Tale of a Tub*'s (1704) "Digression on Madness." For Swift, Bacon was one of the primary originators of "Modern" scholarly methodologies. He was, after all, acclaimed among Moderns for having established "*New Empires by Conquest*" within natural philosophy by "*The Advance and Progress of New Schemes.*"[4] Swift's satires on science have received much scholarly attention, but the absence of critical analysis of Swift's satire of Baconian aphorisms stands as a significant omission within Swift scholarship. It is not only that the empirical methods Swift mocked were many. Swift's engagement with Bacon reveals an early eighteenth-century literary interest in a scientific model of mind that has heretofore been absent from histories of the novel.

Of course, Bacon's use of aphorizing within a scientific context was not intended to be a representation of thought or consciousness. As discussed in the introduction, Bacon argued in *The Advancement of Learning* that knowledge induced was best transmitted to readers through probative methods, which were designed to encourage examination and inspire further testing. Antisystematic aphoristic writing was such a method. Aphorisms, though made of words, would help direct readers away from the error-ridden, epistemological traditions of the past, and toward the thingly particulars that would lead to many as-yet-undiscovered principles of nature.

We know that in 1715 Swift owned a copy of Bacon's *Essays* as well as Gilbert Wats's English translation of Bacon's *De augmentis scientiarum* (1623), the Latin expansion of the 1605 *Advancement of Learning.*[5] Thus Swift knew Bacon's works well and would have associated the philosopher's ideas with empirical techniques practiced across a range of scholarly disciplines. When Swift first parodies Bacon's aphoristic style in *A Tale of a Tub*, he imports along with it a series of concerns about autonomous intellectual labor, a mode of knowledge production that Swift devalues by associating it with physical digestion. It was hardly Bacon's goal to promote such individualistic work. Indeed, he accused his perceived opponents, the Scholastics or "Schoolmen," of that crime. Swift, concerned with what he saw as the general trend toward novelty within Modern learning, adopted Bacon's idea-spinning spider, but used it as a figure for Moderns like Bacon instead. A Modern writer's physical interior—"*the Guts of Modern Brains*"—is now full of modern maxims. Swift's Spider has property in his self and authority over that property, but that collection of dirt and poison (to use Swift's words from *The Battel of the Books*) is hardly something to celebrate. Unlike Behn's personas, who find in maxims addressed to another a sufficient and satisfying expression of the inadequation between exterior and interior, Swift's character-personas slide into an alienating incoherence the moment they begin transcribing maxims. In *Gulliver's Travels*, Swift continues to use maxims ironically to give form to a mode of interiority defined by laborious incoherence. In the process, he contributes to early fiction's dynamic engagement with the representation of inner life, a phenomenon that literary historians still frequently associate with the rise of personal autonomy—the very thing against which Swift railed.

This book has identified a conflict within Bacon's philosophy itself, one that helpfully explains his at times diverging views on human subjectivity and its role in scientific practice and development. On the one hand, Bacon took the mind to be the servant of shifting human passions. Understanding and even perception change depending on a person's individual desires and aversions. This version of Bacon—"Bacon the satirist"—eloquently exposes and even rhetorically exploits human mental distortions. On the other hand, an empiricist Bacon harbors some hope that new scientific procedures will bind men to nature's own pedagogy, enabling them to bypass the perils of individual human subjectivity. This essential, forced apprenticeship to nature would be temporary, ultimately leading to its domination by man. Bacon's belief in the world-changing capacity of experimental science was particularly liable to attack by Swift, who as an orthodox Anglican saw in the new science an abandonment of authorized doctrine for the sake of ambition or caprice. Yet both sides of Bacon attracted Swift's ire. For Swift, Bacon the empiricist was a destroyer of traditional foundations, while Bacon the satirist (and aphorist) was a promoter of a mode of intellectual, internal struggle that could, Swift believed, imperil the thinking subject.[6] Both Bacons are behind Swift's technique of using maxims to depict the inner disorder of speakers—a formal innovation that has something to teach us about the literary environment in which the novel emerged. Swift's outrage was perhaps magnified because he shared something important with Bacon: the conviction that our minds are by nature restless, unable to impose stability within.[7] Both men were skeptical of attempts to repair this instability, though Bacon believed it could be gotten around for the purposes of scientific endeavor. Swift did not.

Satire requires its practitioners to be intimate with their enemies. When a writer takes it upon themselves to eviscerate another writer's style through imitation, they must occupy that style from the inside. Swift was a master of this technique. Through imitative critique he often amplified the unseen epistemological and social investments of his originals. His imitation of Bacon, however, posed some interesting problems. Both Bacon and Swift took for granted certain universal, human cognitive inadequacies. Bacon, however, believed that people could benefit from facing the limits of unassisted human understanding. In so doing, people would be forced to turn away from the self and toward nature. Science can advance once we refuse to be trapped within the ramparts of our own minds. Aphorisms direct us outside the self. Swift's maximizing differs from Bacon's aphorizing in insisting that Bacon's form of "knowledge broken" is itself an expression of the mind's native incoherence, and thus its occurrences should be concealed, not published. And yet Swift does publish his imitations of aphoristic knowledge broken, and in doing so he shifts the purpose of the form, treating it as a representation of the mind's internal and disordered activity. No longer a literary-formal method to remind readers of and direct them toward the not-known in nature, knowledge broken on the page in Swift now figures the mind itself as the seat of not-knowing. Swift considers the possible degrees of ironic play between maxim and fictional context in the *Tale* and *Gulliver's Travels*. In these two works we see him reflecting on his own satirical methods while keeping in mind their consequences for a reading public.

Swift's Baconian impersonations are tightly interwoven with first-person prose that references cogitation. Swift's most thorough engagement with Bacon occurs in *A Tale of a Tub*, a work in which Swift experiments radically with the first-person voice. To us, Bacon's aphorisms read as difficult, unfamiliar, and far from first-person narration. In chapter 1, however, we saw how impersonal, ironic maxims could be brought into contact with a first-person perspective. Behn's goal was to demonstrate for her readers the philosophical authority they could access through an act of personalized rewriting that would put them in touch with a larger, collective force. Swift, by contrast, sees only illiberality in personalized maxims. Departing from Behn's intentions, Swift uses Baconian maxims to extend his experimentation with persona, associating the fictional Author that tells the *Tale*—the voracious writer "of and for GRUB-STREET" who is the source of the insistent "I"—with methods of modern scholarship stretching back to Bacon (40). In doing so, Swift attempts to embarrass the early empiricists with the aid of their own weapons. Bacon designed his use of aphorisms, in part, to inspire humility in readers by drawing attention to the disorder and inadequacy of the undisciplined understanding. Bacon took a risk in combining a satire of intellectual weaknesses with his multidirectional, deliberately disordered instructions for empirical inquiry ("Particulars," he wrote, "being dispersed, do best agree with dispersed directions").[8] Swift's parody of this disorder-embracing form, combined with his own preoccupation with intellectual disorder, brought individual psychology into focus as a problem—and a problem worth exploring—in fiction.

With a fresh sense of the impact of Bacon's prose style on Swift we can begin to see that mental labor can be given form in realist fiction through the employment of sententiae and not only, as is more commonly acknowledged, through techniques of mimetic representation.[9] Swift's first-person accounts raise the question of what is happening under the surface, inside the head—the question of what I've called "inwardness" and others have called "interiority"—without recourse to typical indications of subjectivity: explicit references to personal perceptions, thought, and feelings. By the middle of the eighteenth century, the early psychologically realist novel (the subject of the next chapter) gives a social shape to individual inwardness. It accomplishes this goal with most poignancy when its characters do not speak directly to their experiences or do not intend to do so. Swift's interest in interiority also extends to considerations of psychological privacy and what exists within the confines of the head, considerations that carry us more firmly into the territory of the novel.

A glance at the long career of sententiae in moral philosophy reveals that Swift's experimentation with literary interiority took shape around his critique of realist fiction's ethical pretensions. In *Gulliver's Travels*, maxims become a formal tool for exposing the weaknesses of autonomous individual thought without seeming to posit, in the process, the existence of unimpeachable rationality on earth. As in *A Tale of a Tub*, the primary pedagogical import of the maxims is to check a reader's pride in the "rational" tools that humans need and that a race as purely rational as the Houyhnhnms do not. As in the *Tale*, the maxim does not dictate but rather awakens the reader's awareness of intellectual incoherence. In the *Travels*, Swift takes this a step further, linking such incoherence to moral inadequacy. Swift observed the ineffectiveness of conventional instruction: "I forget," he writes, "whether Advice be among the lost Things which, *Ariosto* says, are to be found in the Moon: That and Time ought to have been there."[10] Swift, it turns out, was like Bacon in that he was willing to take rhetorical risks for the sake of innovative forms of persuasion. Recognizing this makes available for reconsideration the oft-noted incongruity between Swift's ethics and his rhetoric.[11] For, despite holding the orthodox position that what is required of us is obedience to authority, respect for tradition, and reason tempered by common sense, Swift was willing to sacrifice received wisdom for the sake of satiric effect.[12]

This brings us back to Gulliver's inability to access the full meaning of what he has written down. We know more than Gulliver does at that moment. Readers, therefore, must work to perceive the mistake on their own, without a narrator's guidance, and this gives immediacy and life to what would otherwise be dry pedantic prose. The star of the passage is neither Gulliver nor Swift speaking through him, but the mental error ironically disclosed. Swift occupies an unusual spot within the history of English fiction. By focusing our attention on the revelation of personal limitations, Swift is doing something novelistic at the moment of the genre's emergence. But this protonovelistic technique is in the service of a message that is resolutely anti-novel: knowledge originating within the individual is not adequate and coherent but rather alienating and unintelligible.[13]

Empiricism and Aphorism

Swift's adoption of Baconian aphorizing invites us to reconsider dominant accounts of empiricism's relationship to the singular minds of early prose fiction. In *The Rise of the Novel*, Watt does not argue for a direct causal relationship between literary realism and what he calls "philosophical realism," but attributes the development of both to a growing ideology of individualism. To Watt, a celebration of individual perception and action underwrites the novel's "formal realism," its mode of representation that "makes us feel . . . we are in contact not with literature but with the raw materials of life itself as they are momentarily reflected in the minds of the protagonists."[14] As the word "minds" suggests, throughout his account of the close, observational style of early English novelists, Watt stresses their interest in the psychological as well as the social effects of individuality. John Locke's model of the mind as a container of privately owned ideas that are representations of things hovers in the background of Watt's explanation of how literary character, as refigured in eighteenth-century fiction, begins approaching particular identity as a consequence of unfolding consciousness and the habitual association of ideas over time. In his consideration of Richardson's epistolary fictions, for example, Watt finds that characters imagine their private experience as a series of mental representations they can simply transfer into written form in order to examine them at length. Despite the many revisions to Watt's thesis offered by scholars over the years, the notion that empiricism in particular allowed fiction writers to imagine mental states as easily representable in everyday prose remains strong.[15] As Jonathan Kramnick writes, "There is a relation between the empiricist model of thought and the way in which theory of mind problems were developed in the early novel."[16] Locke's own attack on the abuses of language and misleading effects of rhetoric in book 3 of *An Essay Concerning Human Understanding* (1689) is often cited as evidence of the epistemological pretensions of a "plain style" (discussed in the introduction) believed to bring words into a closer alignment with things. The writer simply needed to attend carefully to the parts of and relationships among mental representations, naming each in an appropriate and consistent way.[17]

Yet whereas Locke focuses on furnishing the mind, Bacon insists on our need to renovate a mind already furnished. Despite his faith in the collective power of human knowledge, he was far from believing that a single understanding could naturally perceive the world as it really is. As discussed in this book's introduction, Bacon believed that one of the first steps toward grasping nature's laws involves correcting the mind's tendency to be "hooked by hearsay and debased doctrine, and occupied by thoroughly empty *illusions*."[18] Because traditional modes of reason prevent us from accessing nature's truths, we have to remake logic and, along with it, our methods of presenting information gained. Aphoristic writing—composing one's observations as a series of pithy abstract principles for further testing—could, Bacon believed, be useful in this case. Such disparate observations represented partial knowledge, or knowledge in the process of formation. And so in the *Novum organum,* or "New Organon," Bacon presents his new scientific method, induction, entirely in aphorisms. Aphorisms promote thought while suspending judgment, a state of mind useful for scientific inquiry: "For Aphorisms, except they should be ridiculous, cannot be made but of the pith and heart of sciences: for discourse of illustration is cut off; recitals of examples are cut off; discourse of connexion and order is cut off; descriptions of practice are cut off; so there remaineth nothing to fill the Aphorisms but some good quantity of observation."[19] Although in moral philosophy aphorisms can be used to inspire virtuous action, the lack of system in Bacon's aphoristic writing is particularly useful for the scientific pursuit of new paths of knowledge. What ends up being communicated between writer and reader is system in the process of development—a development that progresses independently of the biases, dispositions, or humors of any individual intellect. "Aphorisms," Bacon wrote, "representing a knowledge broken, do invite men to enquire farther" (*A* 235).[20]

Streams of Sententiousness

What Bacon wrote of induction, his new logic, could easily be applied to *A Tale of a Tub,* the work that launched Swift's literary career, but that he never publicly owned. As Bacon says of induction, "It is not easy to get hold of, it cannot be picked up in passing, it does not flatter intellectual prejudices, it will not adapt itself to the common understanding except in its utility and effects."[21] But if the difficulty of Bacon's presentation of his induction in the *Novum organum* is increased by its aphoristic presentation—its lack of discursive filler—then the difficulty of the *Tale* arises for the opposite reason: it is completely overwhelmed by unnecessary discourse. Although the work ostensibly has two aims distinguished by two different satirical methods—a satire on abuses of religion handled in the allegory of the coats, and a satire on abuses of learning treated in the digressions—the satire on learning predominates, with dedication, preface, introduction, footnotes, digressions, and learned references endlessly delaying and closing in on the little of the story that remains.[22] Taking aim at new, detail-oriented methodologies, which he associated broadly with empiricism (namely, philology as practiced by Richard Bentley, one of his personal targets in the *Tale*), Swift has his Author exclaim that, because everything smart has already been said, the hunt for the textual particular is all that is left. It is the labor of copying particulars that the satirist next ridicules: "What remains therefore," Swift's Author declares, "but that our last Recourse must be had to large *Indexes,* and little *Compendiums; Quotations* must be plentifully gathered, and bookt in Alphabet; To this End, tho' Authors need be little consulted, yet *Criticks,* and *Commentators,* and *Lexicons* carefully must" (97–98).[23] This is the Modern "Method, to become *Scholars* and *Wits,* without the Fatigue of *Reading* or of *Thinking*" (96). Through the absurd claims of his persona, Swift intentionally simplifies and distorts the idea that there is scholarly benefit to gathering textual particulars for use as evidence in learned debate.[24] In Swift's hands, all the groundbreaking new approaches to questions of meaning and being are reduced to the following principle: don't think, just collect.

Of course, Swift is the collector behind this mock-book, a fantasia of learning presented as a monstrous consequence of Bentley's methods of knowledge production. At the heart of the joke is the author-persona, whose manic compositions are made on the spot, monuments to intellectual busyness verging on madness. Through this Author, Swift suggests that if Modern methods of scholarship avoid error by foregrounding evidence and streamlining intellectual procedure—by, as Swift puts it, avoiding "the Fatigue of *Reading* or of *Thinking*"—they simply install a new version of error in the form of self-perpetuating labor with no sense of purpose or relief. Drawing on Bacon, we might say that this labor delays or prevents hasty systematization by giving precedence to parts over wholes. And, indeed, when Swift represents this method of scholarship in the *Tale,* he leans heavily on Bacon's aphoristic form. Entire passages become disordered florilegia, the textual manifestation of delayed judgment (or "not-thinking," as Swift would put it) actively pursued. Take this example from section 7, "*A Digression in Praise of Digressions,*" in which the Author rhapsodizes on his method of composition, which also happens to be a method of attack: "For, the Arts are all in a *flying* March, and therefore more easily subdued by attacking them in the *Rear.* Thus Physicians discover the State of the whole Body, by consulting only what comes from *Behind.* Thus Men catch Knowledge by throwing their *Wit* on the *Posteriors* of a Book, as Boys do Sparrows with flinging *Salt* upon their *Tails*" (96). This is not the end of this stream of "ends." The form of the passage is relentlessly Baconian. What Swift gives us—in a very literal, formal sense—is aphorism after aphorism, or a style mercilessly discontinuous and detached. The thought, if we can call it that, is propelled not by analytic unfolding but by blunt accumulation. The mark of selection (ends abound) only accentuates the passage's lack of clarity and focus, for, to paraphrase Bacon, there is no illustration, no example, no order, no "descriptions of practice." In fact, there is the sense that these are all "cut off." All extraneous rhetorical material has been lopped away.

It is often assumed that Swift's Author is merely a plagiarist or collector of cliché. He works by stealing the work of others, so what we read is not really his. According to these interpretations, the satirical point of a passage such as the above is that original thought decays in an age of mechanically reproduced print, when reader and writer are no longer united in an authentic intellectual exchange.[25] Yet with the help of a Baconian frame one can argue the opposite: that through his Author's streams of sententiousness Swift choreographs an interaction between writer and reader that is all mind, albeit an empty general one. One way Baconian aphorizing inspired thought was by confronting readers with the actuality of their own ignorance with regard to nature: "The subtlety of nature far surpasses the subtlety of sense and intellect, so that men's fine meditations, speculations and endless discussions are quite insane, except that there is no one who notices."[26] In other words, to become actually intellectually and technologically powerful—and to encounter "sanity" in such a reformed condition—we must engage in unknowing (leaving everything we thought we knew behind). In the *Tale,* Swift turns Bacon's play on sanity and insanity against him, assigning the lack of connection between aphorisms to an incoherent writer frantically composing on the spot rather than to a philosopher carefully dividing and distilling truths in the process of a larger project of knowledge advancement. Swift does not mimetically depict intellectual disorder but rather transposes into a fictional context the confrontation with present epistemological ruination that Bacon recommends. We see this again in section 8 of the *Tale* (concerning the imaginary sect of "Æolists," or worshippers of wind), when Swift opts for a rapidity of sententiousness that approaches mania. Here is the Author contemplating why so many cultures believe in both God and the Devil:

> Whether a Tincture of Malice in our Nature, makes us fond of furnishing every bright Idea with its Reverse; Or, whether Reason reflecting upon the Sum of Things, can, like the Sun, serve only to enlighten one half of the Globe, leaving the other half, by Necessity, under Shade and Darkness; Or, whether Fancy, flying up to the imagination of what is Highest

and Best, becomes over-shot, and Spent, and weary, and suddenly falls like a dead Bird of Paradise, to the Ground. Or, whether after all these *Metaphysical* Conjectures, I have not entirely missed the true Reason; The Proposition, however, which hath stood me in so much Circumstance, is altogether true. (103)

Swift's speaker goes on way too long, lost in a minefield of potential explanations. Repetition seems to proceed from compulsion, and every option rises to the surface, only to descend into darkness, death, or misdirection. Rather than progressing, the writer—that insistent "I"—is stuck ruminating, worrying a question compulsively. If one were to remove the "Or, whethers," we would have a series of provocative maxims about human nature.

I may seem to be building a case for there being psychological depth to the *Tale*'s putative Author. But my point is neither about character in the traditional sense nor about the speaker as "a symbol of the imagination's collision with reality," as Ronald Paulson has characterized him.[27] Historians of the novel agree that the elaboration of psychological truth depends on the gradual accumulation of thought and feeling over time. As Frances Ferguson puts it in her essay on *Emma* (1815), individual psychology depends on "the ability to be self-conscious, which is not so much the ability to be accurate about one's own statements and assessments . . . as the ability to see oneself as an interconnected whole."[28] According to some scholars, the epistolary novel was the perfect medium for exploring this extended self-consciousness, because intimate letter writing provided an excuse for the obsessive examination of thoughts and actions—what they were in the past, and how they might change in the future.[29] Although the inwardness on display in the personal letter is already marked by the social world (letters are, after all, addressed to others), it can nonetheless be seen as a document of identity and a record of the tensions inherent in a conscious self that is at once an experiencing subject and an object of reflection. I bring up this account to demonstrate the radically different vision of interiority expressed in these moments of the *Tale*. Swift, I am arguing, adopts Baconian aphorizing as part of his attack on modern scholarly methodologies. In adopting it, however, he had to reckon with Bacon's idiosyncratic use of the form as a means of provoking readers to inquire into peculiar instances of natural phenomena, a provocation designed to turn them away from the present state of mental disrepair and toward scientific advancement.

Interiority as a state of uncomfortable and frequently incoherent struggle is, then, both an idea Swift borrowed from Bacon and the foundation of Swift's satire of Bacon.[30] Literary historian Morris Croll memorably described how seventeenth-century "baroque" or "Anti-Ciceronian" writers like Bacon could portray "not a thought, but a mind thinking.... They knew that an idea separated from the act of experiencing it is not the idea that was experienced. The ardor of its conception in the mind is a necessary part of its truth."[31] As opposed to the more common empirical model of the mind as a mirror in which experience is reflected (and later paired with words that refer to things), the Anti-Ciceronian style imagines the mind as a muscle that struggles in order to know. Swift shared this view of individual intellectual struggle in the face of the not-yet-known as a truth of inner life, but was far less sanguine about the possibility of harnessing it. He certainly did not believe it could be harnessed for profit or "advantage." Further independent inquiry was not what Swift desired. His satire works, then, by using streams of sententiousness to represent disorientation openly and disastrously pursued. He creates the sense that unthinking is unfolding as we read, and he assigns that unthinking to a person by combining these aphoristic passages with first-person prose.

Bacon worried deeply about the human intellect. To him, not just bad habits of thought but some essential fallen-ness in human understanding could thwart even empirically oriented inquirers, sending them in circles, or chaining them to private intellectual obsessions. Yet, ultimately, Bacon believed in the possibility of reform. Swift did not. In the *Tale*, he dwells with the mind's deceptions and, by means of an impressively flexible satirical technique, holds together what seems to us a paradox: the idea that not thinking properly is a consequence of being too inwardly oriented, rather than not inwardly oriented enough.

Private Interiors and the Shame of Public Digestion

In discussions of early realist fiction, the notion of interiority often depends on a spatial (frequently domestic) metaphor to describe a penetrable exterior through which we might reach an interior self. For example, an oblique reference to such a metaphor enables Watt to conflate domestic with psychological privacy when he writes that the "direction" of Richardson's narrative "is towards the delineation of the domestic life and the private experience of the characters who belong to it: the two go together—we get inside their minds as well as inside their houses."[32] Jürgen Habermas argues along similar lines in *The Structural Transformation of the Public Sphere* that early eighteenth-century literature—including the novel—supported the development of modern subjectivity that was already underway thanks to the new privacies conferred by the bourgeois family. The privacy constitutive of Habermas's public begins in "the conjugal family's intimate domain (*Intimsphäre*). Historically, the latter was the source of privateness in the modern sense of a saturated and free interiority."[33] Critics have responded to Habermas with greater sensitivity to the roles of gender and ideology in the construction of the private subject.[34] And such critiques have in turn enriched discussions of the relationship between privacy and literary interiority—discussions that now attend to the role of gender in the eighteenth-century construction of privacy as an ethical issue.[35] But Swift's satire approaches the morality of privacy from a different direction.

In the aphoristic moments in the *Tale*, Swift presents Baconian aphoristic form as if it encourages private intellectual struggle. Swift then works out the social implications of publishing such a private struggle. Bacon himself did not encourage this type of extended, self-absorbed and self-absorbing inner struggle within empirical endeavor. His new induction was designed to help natural philosophers escape from the tyranny of Aristotelian philosophy as well as the tyranny of their own faculties. Our natural intellects, according to Bacon, are characterized by contradictory but equally destructive tendencies: the mind "loves to leap to generalities, so that it can rest; it only takes it a little while to get tired of experience," and yet "the human understanding is ceaselessly active, and cannot stop or rest, and seeks to go further; but in vain."[36] Bacon chose aphorisms because he believed they could defuse the egotistical energy of writers and readers involved in intellectual exchange. In the *Novum organum* his aphorisms were designed to encourage intellectual labor—among many people and across a wide expanse of time—while providing curious readers with plenty of opportunities to rest with stable, "general" observations. In other words, epistemological disrepair is not corrected but is rather harnessed for the sake of the gradual process of knowledge production. And the more public and collective the process is, the better.

Swift takes from Bacon the idea that constant, active production is necessary for the advancement of human knowledge and asks whether the new science can make good on its promise to benefit the public, given the demands of endless inquiry it places on the private individual. In the

Tale, Swift gives this situation concrete figuration. Imagine, he suggests, your own physical digestion: a necessary process that can be uncomfortable and even painful, but the details of which are thankfully unknown to others. Would you want that process made public? Swift answers this question in the "Conclusion" to the *Tale*. There, his speaker claims to have produced the work we have just read merely as a vehicle for the witty remarks he was unable to bring off in public. He had to write to unburden himself of his "laborious Collection of Seven Hundred Thirty Eight *Flowers,* and *shining Hints* of the best *Modern* Authors, digested with great Reading, into my Book of *Common-places*" (136). He indirectly cites the lapsing of the Licensing Act in 1695, the recent "Liberty and Encouragement of the Press" that has allowed him to publish freely—too freely, in fact, for "the *Issues* of my *Observanda* begin to grow too large for the *Receipts*. Therefore, I shall here pause awhile, till I find, by feeling the World's Pulse, and my own, that it will be of absolute Necessity for us both, to resume my Pen" (136). Through the Author, Swift represents self-driven composition as a process of internal labor over impersonal materials (those "Observanda" and "Common-places"). The suggestion of intestinal distress that creeps into the Author's account of his writing crystallizes the connection between interiority and roiling, uncontrollable eruptions that are published and read by others. The turn to the body is typical in satire, as is its use as a mechanism for inspiring shame. But we would be right to remember here that Bacon himself used a physical analogy when discussing his aphoristic method. The *Novum organum* presents "the actual art of interpreting nature and of the true operation of the intellect: not in the form of a regular treatise, but digested, in summary form, into aphorisms."[37] Bringing to the surface certain understated currents in Bacon's analogy, Swift's satire objectifies and externalizes the motions of the mind without divorcing those motions from a sense of privacy and shame.

Modern developments in science and scholarship rather than historical changes in intimate life led Swift to a consideration of the interiority of first-person prose. The Author's private writings are aligned not with freedom from intellectual domination but with that which should be hidden. Needless to say, the whole thing is uncomfortable to read. If the psychologically oriented novel fosters in the reader a desire to penetrate the private thoughts of a character, then Swift is everywhere frustrating this desire by making his narrator disturbingly "open."

Maxims in Realist Fiction

The maxim, then, is the means by which Swift accesses something like novelistic discourse, while simultaneously promoting values that are resolutely anti-novel.[38] I now want to consider what this looks like in *Gulliver's Travels* in order to show how Swift's negotiation of the maxim in this more baldly novelistic work challenges us to recontextualize not only the philosophical but also the ethical underpinnings of represented thought in early fiction.

In the conclusion to his *Travels*, Gulliver confidently asserts that his work will serve "the PUBLICK GOOD"—"for," he writes, "who can read of the Virtues I have mentioned in the glorious *Houyhnhnms*, without being ashamed of his own Vices, when he considers himself as the reasoning, governing Animal of his Country?" As an afterthought he adds: "I shall say nothing of those remote Nations where *Yahoos* preside; amongst which the least corrupted are the *Brobdingnagians*, whose wise Maxims in Morality and Government, it would be our Happiness to observe" (438). Poised to instruct, Gulliver privileges example over rule. And yet ultimately the distinction proves specious. Few Brobdingnagian maxims are actually reported in the *Travels*, and while one Lilliputian rule seems a possible statement of Swift's own beliefs (the "Maxim" that "among People of Quality, a Wife should be always a reasonable and agreeable Companion, because she cannot always be young" [91]), another simply captures strategies of domination notable for their absurdity ("It is a Maxim among these Lawyers, that whatever hath been done before, may legally be done again" [370]). In his sojourn in Houyhnhnmland, however, Gulliver finds a maxim he can wholeheartedly endorse. The Houyhnhnms have one—and only one—"grand Maxim": "to cultivate *Reason,* and to be wholly governed by it." In a final flourish Gulliver adds: "Neither is *Reason* among them a Point problematical as with us, where Men can argue with Plausibility on both Sides of a Question; but strikes you with immediate Conviction; as it must needs do where it is not mingled, obscured, or discolored by Passion and Interest" (401–2).

Swift's superbly rational horses lack the mental weakness that makes virtuous living a matter of constant negotiation. For this reason, Houyhnhnm life is devoid of two things: psychological struggle and the precepts of practical morality. Houyhnhnms need only one universal rule, because they live a life of detached, rational tranquility—the life, that is, of a maxim, and not of the person who uses it. Passion and interest are what made maxims necessary to the ancients who used them in the "spiritual exercises" that Pierre Hadot has identified as part of a philosophy of the self.[39] For Stoics as well as Epicureans, the possession of universal truths capable of overturning subjective bias constituted part of a therapy that could "raise[] the individual from an inauthentic condition of life, darkened by unconsciousness and harassed by worry, to an authentic state of life, in which he attains self-consciousness, an exact vision of the world, inner peace, and freedom."[40] This perfect consciousness was deeply impersonal, requiring identification with an eternal order existing outside the self.

This brief summary suggests that moral philosophy might be an adequate frame for reading the maxims in *Gulliver's Travels*. And yet Swift was no believer in Stoic self-command, which he critiqued, ironically, in aphoristic form: "The Stoical Scheme of supplying our Wants, by lopping off our Desires; is like cutting off our Feet when we want Shoes."[41] (Desire is central to human being and experience, and, as such, it is a burden that defines us.) What to make, then, of Gulliver's commitment to maxims?

In *Gulliver's Travels*, maxims become a formal tool for exposing the weaknesses of individual thought without seeming to posit, in the process, the existence of unimpeachable rationality on earth. As in *A Tale of a Tub*, the primary pedagogical import of the maxims is to check a reader's pride in the "rational" tools that humans need and that a race as purely rational as the Houyhnhnms do not. As in the *Tale*, the maxim does not dictate but rather awakens the reader's awareness of intellectual incoherence. In the *Travels*, Swift takes this a step further, linking such incoherence to moral inadequacy. For example, in the fourth book of the *Travels*, in a passage discussed in this book's introduction, Gulliver gives this account of his domestic economy in Houyhnhnmland: "I soaled my Shoes with Wood which I cut from a Tree, and fitted to the upper Leather, and when this was worn out, I supplied it with the Skins of *Yahoos*, dried in the Sun. I often got Honey out of hollow Trees, which I mingled with Water, or eat it with my Bread. No Man could more verify the Truth of these two Maxims, *That, Nature is very easily satisfied;* and, *That, Necessity is the Mother of Invention*" (416–17). Swift provides us with a recognizable prose structure: experiential narrative interpreted according to general laws. These general laws then double as moral lessons for readers who may have missed the example, as Gulliver suggests in his conclusion. Things change on a closer look, however, when we realize that the utterly familiar maxims justify a form of cannibalism: in want of shoes, Gulliver removes and apparently cuts and sews Yahoo—or human—skin. Experience in isolation has warped Gulliver's humanity and distorted the methods by which he hopes to communicate his experiential knowledge to others. Despite their structure his sentences do not convey neat moral closure. Rather, the maxims' presence conveys their lack of moral coherence.

I have been arguing throughout this chapter that when we find maxims in Swift's fictions we should be on the lookout for inwardness effects. Is there room for finding them here? Scholars of English fiction largely agree that one of the best methods for giving voice to a character's consciousness was, if not "invented" in the late eighteenth century, only used inconsistently and rarely before that time. I am speaking, of course, of free indirect discourse.[42] I want to suggest now that by an ironic deployment of maxims, Swift achieves a similar effect much earlier. The above passage's peculiar irony has to do with the way the fictional prose travels among multiple perspectives. In play we have: Gulliver's perspective; the perspective of the maxim (something like the viewpoint of law itself); the "notional" perspective of the reader, invoked by maxims that temporarily compel us to recognize them as true; and, finally, the more shadowy perspective of the satirist pulling the strings. It is an indeterminate mixture that puts serious pressure on our ability to identify or disidentify with Gulliver through his thought process. Claude Rawson, in a discussion of the role of Augustan satirical conventions in Jane Austen's work, describes the operation of free indirect discourse in terms that emphasize such abrupt changes in moral viewpoint: "It combines the ostensibly factual reporting of speech and thought with complex and shifting intimations of judgmental perspective: of the attitudes or point of view, for example, not only of a first- or second-hand reporter, or of a narrator (whether 'personalised' or 'authorial'), but also of participants in the reported conversation, and even those of the notional reader."[43] Austen clearly engages in such a complicated dance of perspectives, as a brief example from *Emma* will illustrate. The narrator enters Emma's consciousness just after a conversation in which Mr. Knightley accuses Emma of a misapplication of her reason: "He had frightened her a little about Mr. Elton; but when she considered that Mr. Knightley could not have observed him as she had done, neither with the interest, nor (she must be allowed to tell herself, in spite of Mr. Knightley's pretensions) with the skill of such an observer on such a question as herself, that he had spoken it hastily and in anger, she was able to believe, that he had rather said what he wished resentfully to be true, than what he knew anything about."[44] In much of this passage the narrator uses conventions of indirect speech to report Emma's thought. The narrator is close to Emma, but there remains a division between their two entities. It is only between the parentheses that we feel we have gained access to the very movement, the very sound, and the very wrongness of Emma's own thought. The parentheses—much like the italicized maxims in the above passage from *Gulliver's Travels*—focus our attention on the way that "Passion and Interest" distort Emma's thinking. Although she is blind to this distortion, we are not. Our intimacy with her error generates the sense that we might now avoid the same. Or, at least, Austen's fiction—through Emma's narrative of moral development—suggests the possibility of reformation. Swift's use of the maxim offers no such possibility.

There are, of course, important differences between free indirect discourse and Swift's technique. Most importantly, Swift is working within first-person rather than third-person prose. Thus Swift's passage contains none of the sentences Ann Banfield calls "unspeakable" because of their grammatical combination of the first and third person.[45] And yet Swift's passage also shows more about Gulliver than it tells, though it does so by giving us a glimpse not of a coherent self, but of a distressingly borderless incoherence. Gulliver makes his ideas public, but there is a depth to them, because we realize what he is thinking only on a second read. Michael McKeon examines the production of a similar illusion of depth when he advances his own account of free indirect discourse. According to McKeon, shifting narrative perspectives open a formal space for subjectivity to fill:

> As a method of internalization, free indirect discourse does not, strictly speaking, reach a "deeper" level of consciousness in characters than that already accessible through first-person narration (whether epistolary or autobiographical) and third-person "omniscience." Rather, the effect of greater interiority is achieved by the oscillation or differential *between* the perspectives of narrator and character, by the process of moving back and forth between "outside" and "inside," a movement that seems palpably to carve out a space of subjective interiority precisely through its narrative objectification.[46]

There is perspectival oscillation in the maxim passage from the *Travels* analyzed above, an oscillation that opens a space of "subjective interiority." And yet, because Swift deals in first-person prose rather than omniscient narration, what constitutes "outside" and "inside" in the passage is far less clear. As opposed to the "space" of subjectivity that Austen makes accessible through free indirect discourse, the "space" of Swiftian subjectivity operates as a trap for the reader. It is not something a reader can easily enjoy or freely exploit on behalf of their own moral development. In Swift, we are exposed to a distressingly incoherent interiority from which there is no "outside"—no escape.

Jenny Davidson can help us make sense of this strange combination of impersonal and personal—or public and private—qualities in the maximic passage from the *Travels*. Swift's major author-narrators, she argues, "go well beyond the conventionally understood limits of personal identity, with its criteria of coherence and consistency."[47] Neither with the *Tale*'s speaker nor with Gulliver is Swift ironically violating secure boundaries of personal identity. Rather, through the canny manipulation of maxims, he challenges those boundaries as they come into being.[48]

Reintroducing Bacon can help us further understand the complex dynamic Swift sets up among character-narrator, author, and reader by playing with these boundaries. We might say that in *Gulliver's Travels*, "knowledge broken" appears as the narrating subject's blindness to his own unreason. Such self-blindness is jarring for a reader, who by reading engages in the penetration of a mind but receives little pleasure in doing so. For both superior insight and sympathetic identification are impossible. In this moment of maximizing, which is also a moment of uncomfortable revelation about Gulliver, Swift's satire reaches an ethical crisis. According to Gulliver, in his Houyhnhnm paradise, one life—one man—is all it takes for moral law to exist: "No man could more verify the Truth of these two Maxims." The other side of this is that, through the act of verifying his own maxims, Gulliver becomes "no man."

Beginning from Locke, accounts of eighteenth-century literary interiority find private thought amenable to discursive, narrative elaboration. In this chapter I have taken a different approach to eighteenth-century literature's interest in private thought by examining a tradition in which the personal and social benefits of inwardness are not so clear. Reading Swift through Bacon helps us see sententious instability as part of a fictional experiment in representing the limits of individual consciousness that does not assume the moral productivity of an intimacy with internal incoherence and error. Swift's characters have and record experiences, but they are not self-conscious. As readers we have come to expect the novelistic illusion of mind to depend on reflection. Self-consciousness is the source of inwardness's apparent depth. One of the great ironies of the place of Swift's fiction in literary history is that the excessively personal writing it denigrates was soon to prove a sensation in novels such as *Clarissa* (1747–48). Not even Swift's vision of deranged first persons could stem the flood.

3

The New Realism of Literary Generalization in Richardson's *Clarissa*

AFTER HE RAPES Clarissa Harlowe, Robert Lovelace undergoes what he calls his "trial." The proceedings take place in his family's drawing room, where a collection of relatives present him with a letter Clarissa has written to his aunt Lady Betty Lawrance. Readers have already encountered this letter, in which Clarissa writes plainly: "Again made a prisoner, I was first robbed of my senses; and then (why should I seek to conceal that disgrace from others which I cannot hide from myself?) of my honour."[1] When his relatives suggest that he might repair this injury through marriage, Lovelace accuses Clarissa of reckless publicity. He complains that she "is everywhere, no doubt . . . pursuing that maxim, peculiar to herself (*and let me tell you, so it ought to be*), that what she cannot conceal from herself, she will publish to all the world" (1035). Clarissa's use of parenthesis signals an act of reflection and even self-doubt amid her statement of fact. It is a moment of the kind of reflective self-consciousness of which Swift's Hack is incapable. Here, Clarissa says she was raped while at the same time acknowledging that many would prefer her to conceal this truth. Lovelace would certainly prefer it and communicates as much in his calculated alteration of her original statement. While retaining Clarissa's reference to the purposes of concealment, he reverses the relationship between surface and depth created by her use of parentheses. The question regarding conduct contained within parentheses in Clarissa's sentence becomes a certain maxim in Lovelace's version.

Lovelace, amateur psychologist, has always been interested in how "[Clarissa's] mind works," most especially in its "whimsical" ways (*C* 889). He takes her letters as sketches of an interior from which one might draw conclusions regarding desire, intention, and action. But in his formulation of her "maxim" we find no luxurious description of feeling, no hint of appreciation for depth or nobility of mind. Lovelace selects a form demonstrative of what he takes to be the impersonal flatness of her character following her rape: the maxim eliminates Clarissa's privacy, moving everything that was "inside" (mental states, experience) to the "outside." The maxim is now Clarissa, and Clarissa the maxim: general and widespread *things* from which Lovelace hopes his family audience will turn away, because thoughtless things do not encourage sympathetic attachment as persons do.

Up to this point in this book I have focused primarily on the role of the maxim within philosophy, particularly natural and moral philosophy, and the fictions and fictional minds inspired by it. The ground shifts, however, as these disconnected, empirical maxims evolve alongside early novelistic fiction. For Behn, maxim writing—especially personalizing general maxims for a particular recipient—could be a practice of philosophical inquiry augmented by elements of amatory secret history. By inappropriately personalizing general maxims and giving them her own disordered arrangement, she performs a literary embrace of the "fleshly intellect," which she portrays as feminine, socially unisolated, and physically liberated. Swift shares the view that writing in disordered maxims portrays the fleshliness of intellect, but his satires urge caution around such painful and shameful public displays. Yet through his satires Swift carried into prose fiction concerns about the stakes of individual inquiry. He also maintained the physicality or materiality of cognition *and* noncognition. Swift's satires suggest that we should pray for a miraculous return (and it would be a miracle) to an intellectual era prior to the one that has plunged us into the isolating, individual horrors of unprocessed, laborious incoherence. In this sense, Swift is not unlike Lovelace, who yearns to reconceal what Clarissa has opened and to reprivatize what she has published. *Clarissa* is not Richardson's first novel, and by the time he wrote it he was aware of a paradox at the heart of the early novelistic representation of interiority. The cause of this paradox was the need for readers to accept novelistic writing as capable of representing private experience meaningfully to a wide range of potential readers. Ironically, the best way to accomplish this was by eliminating the signs of unique, private experience altogether by eliminating its disordered, disjunctive qualities. *Clarissa,* however, only adheres to this unspoken literary agreement to a limited extent and deploys disordered and inadmissible literary generalizations at the crux of the novel's tragedy to remind readers that their disordered, disjunctive experiences are both unique and nonunique from the experiences of others. Literary generalizations are furthermore particularly able to inspire readers to see and make new arrangements and relationships as part of their experience of the world.

Of course, alongside the empirically inflected, disordered maxims that I have so far charted throughout this book, more traditional moral maxims continued to flourish in eighteenth-century fiction. By the time Richardson wrote *Clarissa* (1747–48), maxims were integral to the eighteenth-century novel, with the two forms evolving in relation to one another and alongside modern programs of information organization. Richardson went so far as to extract his own maxims from *Clarissa* collating and appending them to the 1751 third edition as *A Collection of Such of the Moral and Instructive Sentiments, Contained in the Preceding History, As are Presumed to be of General Use and Service.*[2] Less often noted in histories of these conjoined forms, however, is that eighteenth-century writers defended the instructional value of new fictional histories by pitting their recognizable worlds against maxims. These defenders assumed maxims had a didactic, instructional function. Thus they contrasted the maxim's epistemological nullity with the vicarious witnessing made possible by plausible fictional worlds. Diderot writes, for example, that while Richardson's fiction moves him, the maxim does not "itself impress any perceptible image upon our minds."[3] Diderot's framework is Lockean; he sees knowledge as property of the individual mind, property readers might gain from books instead of purely from experience, so long as writers agree to deploy consistently the same words to denote the same ideas of things. Samuel Johnson concurred that the new fictional histories evoked in readers a unique sensory response akin to the responses we have to real objects and actions in the world. Maxims, by contrast, teach nothing because we feel and experience nothing when we read them. Johnson writes, "never yet within my reach of observation" has a familiar, reused sentence such as *Life is short* "left any impression upon the mind."[4] We now have twenty-first-century literary cognitivist accounts of how realist fiction produces knowledge, and yet still these accounts of realism's efficacy hew closely to mid-eighteenth-century writers' celebrations of empirical plausibility. "Fictional narration is realist," according to William Warner, "when conducted in such [a] way that it incites in the reader or audience an experience that appears true to reality."[5] The internal experience is identical, whether inspired by active life or by the page.

As compared to these eighteenth- and twenty-first-century accounts of realism, accounts of maxims in early novels suggest that the form channels the *unreal*. Maxims convey nothing, or "no-thing," within Diderot's framework, for they leave no "perceptible image" on the mind.

But what happens to this empirically minded version of realism when characters' and readers' experiences of what is true to reality differ? After Lovelace rapes her, for example, Clarissa's letters advance an account of rape that to Lovelace seems nonreal. He characterizes her communications on the topic as quixotic by turning them into a maxim to which she is chained ("What she cannot conceal from herself, she will publish to all the world"). Clarissa and Lovelace disagree about whether privileged access to Clarissa's interiority is necessary to make certain what really happened. The question this raises for readers of the novel is whether establishing reality with justice depends on the author making a character's interiority intelligible for a reader.

If realism is one pillar of the novel dated to the eighteenth century, interiority is the other. The two are related in the aforementioned defenses of realism, in that the reality effect of these fictions depends on the presence of individualized minds—most pressingly, the minds of "readers"—in which the experience "that appears true to reality" can be realized. Here again the maxim proves a useful foil for realist fictional work. Realism modeled on early empirical protocols requires individuals to produce its knowledge. By contrast, maxims are relentlessly nonindividualistic. Stephen Greenblatt, for example, writing on *The Jew of Malta*, describes Barabas's "proverbs" as "the compressed ideological wealth of society, the money of the mind."[6] With increased use, these anonymous "ideological riches" "render [Barabas] more and more typical [and thus] *de-individualize* him."[7] I have said that maxim and novel evolve alongside one another, but how can maxims contribute to the early novel's realism if they thwart its construction of modern personhood? Indeed, de-individualization in *Clarissa* would seem to harm the case for the damage done to Richardson's heroine, which rests on the violation of her rights as a self-determining person. And because those rights themselves depend on agentive inwardness (Clarissa's capacity for reflection), the maxim Lovelace assigns her seems designed to deny her those rights by denying her interiority.

Yet neither the maxim's generality nor its impersonality denies it real, material existence. As a contained textual object, the maxim has an ontological status akin to the physical copy of *Clarissa* we readers hold. Lovelace fearfully highlights the maxim's accessibility and easy reproducibility when he uses the form to denounce Clarissa's decision to inform his family of her rape. Ironically, in a letter to her daughter on the topic of Harriet Byron in *Sir Charles Grandison*, Lady Mary Wortley Montagu reproduces Lovelace's "maxim of Clarissa" alongside his disgust with Clarissa's reference to her violated body: "Her whole behaviour, which he designs to be exemplary, is equally blamable and ridiculous. She follows the maxim of Clarissa, of declaring all she thinks to all the people she sees, without reflecting that in this mortal state of imperfection fig-leaves are as necessary for our minds as our bodies, and 'tis as indecent to show all we think, as all we have."[8] The "maxim of Clarissa" is directive ("Publish what you cannot conceal from yourself"; "Declare all you think to all the people you see"). It resists the doctrine of sexual shame and its corollary insistence that a woman's body and mind must never become general, widespread, or uninteresting. And the maxim itself, textually reproducible, is both more general and more materially assertive than Clarissa, the exemplary modest woman, is permitted to be. Clarissa's maxim indexes a reality to which her experience contributes, but which is ultimately separate from and greater than her.

What is the status of such a claim in this chapter, centered as it is on a novel that treats the systemic culture of rape?[9] Modern literary critics fear (and Lovelace believes) that the maxim—a form ostensibly representing self-evident truth or socially consented to fact—snuffs out or suppresses private experience. Lovelace's translation of Clarissa's rhetorical question (*I know; why shouldn't they know too?*) into a maxim (*If you can't hide it, reveal it*) aims to suggest that others could learn from her dangerously self-sabotaging behavior, which thus should be stopped, but his translation avoids reckoning with Clarissa's voluntary resignation of her privacy. Indeed, the conflict between Clarissa's question and Lovelace's version of it as a maxim actually clarifies a paradox at the center of private experience as it is figured in the early psychological novel: the idea that individual experience is epistemologically valuable only when it consists of objects of perception organized by reason, and yet that the same experience is most personally powerful and meaningful when it is formless and inarticulable. In giving form to a way through this paradox, literary generalizations in *Clarissa*—from maxims such as the above to general literary quotations applied to an individual's life—suggest that language that circulates publicly and that is, as a free object, easily appropriated can nonetheless remain charged with the possibility of bearing private experience.

The presence in *Clarissa* of maxims that deprivatize and deindividualize knowledge (including partial knowledge or the lack thereof) drawn from experience opens the possibility of a realism no longer linked to contained or constrained, individual psychological interiors or to empirical knowledge accessible only to those who meet the requirements of modest witnessing or neutral observers.[10] So far I have mentioned only one maxim, marked by Lovelace's creative flourishes. That first maxim is, however, representative of a larger category of literary generalization in *Clarissa*, under which we find the following: the "maxim" that Richardson has Lovelace compose for Clarissa, and others like it; the literary fragments comprising the tenth "mad letter"; and brief literary quotations extracted from their context and used by fictional characters to predict effects within the fictional world and beyond. These literary generalizations are neither artifacts of conservative dogma nor object lessons on the novel's comparative freedom from literary-formal constraint. Instead, they are part of *Clarissa*'s interest in a realism grounded in a common material world of which humans are only a part, as opposed to a realism grounded in particularized perceptual experience capable of undergoing a transformation into knowledge only within an individual mind. Such literary generalizations turn readers toward a "real life" that is shared, inequitable, and unredeemable through sympathetic engagement or through an individual's cultivation of meaning within. This is a very different realism than the still-influential formal realism of the early novel that Ian Watt introduced many decades ago. For Watt, realism as an eighteenth-century aesthetic was indebted to the new experimental philosophy's approach to reality as a material world that science makes speak and that necessarily propels human progress. By contrast, the reality at which the realism of literary generalization aims is the assemblage—the set of associations—of which we are a part and we affect, but that is never entirely limited to or controlled by us.

In Defense of the Maxim

By insisting on readers' disinclination toward maxims and attraction to plausible fictions, Johnson and Diderot outlined their theory of the new fictional narratives: examples designed to create an experience of real life will direct moral judgment better than precept, because examples can convey the experience of the real through a descriptive closeness to persons and things. Maxims make no impression on the mind. By contrast, "a character in action is seen; we put ourselves in his place or at his side; we take sides either for or against him."[11] Confronted with modestly edited (or narrated) plausible events unfolding in a recognizable world, young persons with "minds unfurnished with ideas, and therefore easily susceptible of impressions" will watch, be impressed, and draw inferences regarding causality.[12] The reader's belief in the narrated facts is warranted because of the modesty of the author's invention. This experience transmitted through reading can lead to new knowledge for the reader and, ideally, changes in action, because plausible events convey not only knowledge of what the world contains, but explanations of why things happen. Scholars of the early novel tend to refer to such narratives as "realist" because they adapt literary tools to model the epistemic virtues of empiricism, including the referentiality of words to ideas and ideas to the perceptible qualities of things. The following claim by John Bender is paradigmatic: "In point of both thematic exposition and narrative strategy, these novels force readers into the position of neutral observers arriving, probabilistically, at judgments based upon the weight of available facts and reasonable inferences."[13] The plot too guides readers through the evaluation of these "surrogate observation[s]."[14] Readers are the witnesses—the "neutral observers"—of unfolding actions in a plausible albeit fictional world, the consequences of which they can better infer because of their neutrality.[15] The experience of impartial neutrality aside, as Catherine Gallagher reminds us, the data comes from elsewhere—comes preordered and packaged, "shap[ed] . . . through [the author's] fabrication of particulars."[16] The novel's attention to the minute particular suggests reference to the real in order to conjure a believable world that does not solicit belief. Although not inductive in any technical sense, the process of rising to general truths from particulars in a narrative is inspired by inductive practices.

Helen Thompson has compelling argued, however, that a scholarly misunderstanding of Locke's empirical epistemology has led to a further misrepresentation of the relationship between object and observer in the creation of empirical knowledge within early novelistic realism. Using a familiar argument by Michael McKeon as an example, Thompson demonstrates that McKeon's account of novelistic realism depends on the portrayal of Locke's epistemology as follows: a legitimate empirical truth claim requires separation of perceiver from thing perceived, or of subject from object. Instead, Thompson reminds us, Locke "adopts Boyle's corpuscular doctrine of qualities."[17] This eighteenth-century corpuscular "chymistry" as developed by Robert Boyle takes as the empirical ground for truth "not separation but relation."[18] In making her case Thompson departs from Shapin and Schaffer's account of Boyle's experimental philosophy and the production of scientific fact. Boyle is, Thompson argues, much more invested in a theory of imperceptible "corpuscular causes" than *Leviathan and the Air-Pump* allows him to be.[19] To explain how both Boyle's and Locke's empirical knowledge is relational and *qualitative* (as opposed to purely quantitative), Thompson reevaluates Locke's infamous "secondary qualities," arguing that they are "not free-floating ideas but productive powers rooted in matter."[20] Novels such as Richardson's *Clarissa*, Thompson argues, use form—"project the forms, relations, and powers through which empirical apprehension of reality happens"—to know this same "qualitative reality" that eighteenth-century chemistry and empiricism more broadly aimed also to know.[21] *Clarissa*'s realism is not, then, a primarily mimetic attempt to find a prose style that could reflect the world as it is recreated only in individual, circumscribed sensoria.

Thompson's account usefully critiques the "spectator theory of knowledge" as phrased by Ian Hacking and advanced by Shapin and Schaffer's account of Boyle's experimental production of matters of fact. Unlike Thompson's work, this book is more interested in realism's connection to practices of empirical *unknowing* rather than to empirical knowing. Yet *Fictional Matter* remains helpful to my own argument to the extent that it directs us to the primacy and problems of a dualism in the theory of novelistic realism that is clear already in Watt's *The Rise of the Novel,* and that persists throughout the work of more recent historians of the novel such as McKeon and Bender. Thompson writes that "Watt's overarching concern is an individual whose nascent perceptual autonomy propels his or her break from traditional social order. Watt aligns this individual's 'problematic' separation from the collective with 'dualism' that, according to his Cartesian rubric, insulates things 'discovered . . . through the senses' from 'internal . . . consciousness' of the self. For Watt, realist reference to external objects and realist reference to inner selves runs on parallel but alienated narrative tracks."[22] Diderot's and Johnson's accounts—and also Bender's—make it seem impossible for maxims to work in conjunction with the project of fictional realism, because they take that project to be devoted to creating the perceptual experience in readers of rising gradually and inductively from observed particulars to general truths about nature. Words mediate particulars; beyond this they have no place. Alternatively, as Watt's dualism suggests, the words of novelistic realism might refer to the particulars of an inner psychological experience that remains isolated from the perceptual experience of external objects. Because of their generality, maxims do not refer realistically to external objects, and, according to Johnson and Diderot, neither can they refer realistically to internal perceptions or impressions.[23] Maxims that flaunt their form in early realist fiction (indeed, in *Clarissa*) are thus marked as suspect, because they are ostensibly relics of epistemologies that found truth in words and abstractions, not in things full of individual peculiarities.[24] Yet just as Thompson recovers in Boyle's corpuscular chymistry an openness to the causal importance of insensible particulars to empirical knowledge, so I find in *Clarissa*'s literary generalizations an openness to a disordered collection of unimpressive maxims that we can trace back to Bacon's probative method of knowledge broken.

In *Clarissa* characters deal plentifully in book learning—ancient and modern—in the form of maxims. As discussed in the previous chapter, in many ways Swift's *Tale* is a satirical artifact of improper book learning. Yet the readers and writers attacked as bad book learners in the *Tale* are those struggling in garrets to survive, whose writing is part of an embodiment both physical and mental. *Clarissa* targets different bad readers. The most obvious maximizers in *Clarissa* whose sentences are associated with knowledge from books are men in power (at least temporarily), whose reliance on maxims denotes an absence of the insight they would gain by looking about them and actually seeing others. Lord M. never appears in the text but among what Lovelace calls his "*wisdom of nations*" (663). Parson Brand, peddler of Latin sententiae, fails as a Christian because he cannot see past his pedantry.[25] It seems at first that Richardson's dismissal of these men's wisdom lies in its alignment of so-called knowledge and status: this knowledge derived from books is the property of the few. Richardson, however, also presents their maximic knowledge as socially inoperative nonknowledge.

This nonknowledge he associates with interiority. Despite the radically different social positions of Lord M. and the Author of Swift's *Tale*, here we have some overlap between Swift's disordered and disorienting interiors populated with extracts from reading and Richardson's thoughtless repeaters of wisdom. Shall we assume, then, that some of the literary generalizations in *Clarissa* are targets of a distinctly class-conscious attack on a kind of late-stage Scholasticism? Such an argument follows closely the account of realism's origins in empiricism discussed in the beginning of this section. According to an early English empirical tradition, the Scholastics or "Schoolmen" whom Francis Bacon loved to attack were locked up in the axioms (a.k.a. the maxims) of Aristotle "their dictator." These arachnid philosophers, "knowing little history, either of nature or time; did out of no great quantity of matter, and infinite agitation of wit, spin out unto us those laborious webs of learning which are extant in their books."[26] The webs the Schoolmen spin are the product of dark and dusty interiors and represent "the wit and mind of man . . . [working] on itself."[27] The empirical origin story that traces the early realist novel back to Bacon's rejection of Scholasticism's arachnid philosophers is thus more complicated than it first appears, for, according to Bacon, Aristotelian axioms create not knowledge but nonknowledge in the form of an endless generation of caged and encaging human interiority. If, as Lukács would have it, the novel is the modern literary genre that renders philosophy dramatic by emplotting the "adventure of interiority," then the Schoolmen were novelists, not philosophers, though Bacon would hold that they had few real adventures.[28] Here then we see how in a non-Lockean British empirical tradition the efflorescence of interiority does not align with empiricism's commitment to learning from things, from "nature" as separate from the social or political. Brand's and Lord M.'s generalizations are not as antithetical to inwardness as critics generally take them to be. These men are divorced from the real *because* they are all mind.

Within *Clarissa* is a vision of the material world that is skeptical of interiority's unique purchase on truth, but that also refuses to authenticate one privileged version of reality at the expense of other experiences of the real. Although the alternative realism I find in *Clarissa* is not "new," I mark it as such to bring it into contact with what has been called the "new materialism." In one version of the new materialism, Bruno Latour has called for compositionism, a new materialist—but more appropriately, a "new realist"—challenge to the noncompositionist realism demanded by what Latour calls the "Modern Constitution," or the "bifurcation" of human and nonhuman, politics and nature, that Latour traces to the late seventeenth century. For Latour, the scientific reductionist "cage of nature" is one type of composition of human and nonhuman, a "fully *political* way of distributing power." It is "(or rather, it was during the short modern parenthesis) a way of organizing the division . . . between appearances and reality, subjectivity and objectivity, history and immutability."[29] The problem with compositionism (social constructivism by another name) is that things avowedly constructed became vulnerable to the attack of unreality. We are not dealing here with the charge of unreality made against something that is perceived to be "all mind." Rather, we are dealing with a charge of unreality laid at the door of a composition or arrangement that eschews the classic dualistic division between inside and outside, body and mind, self and other. It is this charge of unreality, which in *Clarissa* extends to the fact of rape, to which Richardson's realism of literary generalization responds.

Rape's Forms, or Composing Rape with Paper X

Frances Ferguson's 1987 essay "Rape and the Rise of the Novel" demonstrates that it is not particularized narrative or a "plain style" that justifies the categorization of *Clarissa* as one of the first realist novels. In fact, the novel's private, particularized testimony derives its credibility not from its close reference to "nature," but from the stipulations of Richardson's "to the moment" epistolary technique. The conventions of epistolarity stipulate that the printed page is *actually* a page of hastily handwritten records. According to Ferguson, what actually makes *Clarissa* a realist novel—indeed, a *psychologically* realist one—is its commitment to interrogating the stipulations not only of epistolarity but of the legal "form" or composition of rape itself. If "the psychological novel arises to demonstrate the superiority of individual perception to the world of social forms" then in Richardson's second novel, as Ian Watt argued, "rape becomes the vehicle for the contrast between what could be said in public and proved and what is said in private and believed."[30] Ferguson's argument ultimately departs from Watt's claim that Richardson's novel suggests that the "real truth" is only available in its collection of private correspondence. By contrast, Ferguson argues that *Clarissa* activates the contradiction inherent in the law's formal definition (a.k.a. construction) of "real" rape according to which private meaning is eradicated by the requirements of the (social) form. For the sake of a clear, unassailable judgment, the particular experience of the individual is rendered unnecessary to the determination of whether an act was or was not rape. The law regarding rape was constructed such that, for an act to be rape, both nonconsent and consent must be rendered impossible. To put this in Ferguson's terms, under the law, reality and unreality become tightly wound together, such that "the *reality* of the story of rape" depends on "the relative *unreality* of the victim before society or the world."[31] After explaining this logic, Ferguson argues that Richardson uses this "self-canceling" mechanism within the legal form of rape to interrogate the operation of other social forms that depend for their authenticity on their eradication of private meaning— other forms such as the psychologically realist novel.[32]

Ferguson attributes this commitment to internal contradictions in Richardson to a new "aesthetic": a "mimesis of distinction rather than of similarity, a pitting of stipulation against its internal self-contradiction."[33] Within this new aesthetic, as Ferguson describes it, "the business of forms"—the act of giving shape to reality by social or legal agreement, from the outside in—presents itself as "unfinished."[34] Ferguson's point in using the apparent oxymoron "mimesis of distinction" is to suggest that what Richardson's work imitates is the incompleteness or even the failure of representation—a failure that is very real regardless of its discursive status. No subject, whether fictional or real, can self-represent in such a way that would be comprehensible to others. A knowledge of another's interior depends upon our collective assent to conventions of shared representation. And yet the moment we move toward convention we also move away from truth. Sandra Macpherson has explained that within Ferguson's argument, "psychological complexity" arises from this tension between private meaning and publicly recognized forms that we read for meaning. Ultimately, the "mimesis of distinction" Richardson pioneers is a way to protect a person's "capacity to experience themselves as distinct from the way they are represented by others."[35] Further building off of Ferguson and Macpherson, Wendy Anne Lee argues that the psychological novel is "a self-cancelling form, whose double-purpose is both to construct and to eradicate inner life."[36]

For Ferguson, the "disorderly" literary generalizations in the "mad letters" at the center of the novel make visible *Clarissa*'s self-canceling form. When Richardson's heroine gradually comes back to consciousness three days after Robert Lovelace drugs and rapes her, she immediately begins to write fragments of unfinished letters. Clarissa, we are told, "tears, and throws" them "under the table, either as not knowing what she does, or disliking it" (*C* 889). Richardson provides an explanation for how such documents make their way into a letter and, ultimately, under a novel reader's perusing eyes: Clarissa's rapist cannot bear to transcribe the papers, so the maid Dorcas copies them instead. "Paper X," the tenth and final of these documents, consists of a collection of literary quotations remembered (and, we are to imagine, arranged by Dorcas in the positions reproduced on the printed page). Several of these quotations appear as tilted lines of print in the margins, carrying readers past the mimetic breaking point, violating the accepted convention of epistolarity by which we are to interpret printed characters as hastily handwritten notes.[37] The literary scraps are distinctly maximic in their generality and detachment from the particularities of first-person narration. Take, for example: "When honour's lost, 'tis a relief to die: / Death's but a sure retreat from infamy" and "For life can never be sincerely blest. / Heaven punishes the *Bad*, and proves the *Best*" (893). These literary generalizations come from a variety of tragic dramas, but also from satire, both professional and political.[38] According to the editors of the abridged *Clarissa*, "Much of this paper consists of scraps of poetry that in her delirious anguish Clarissa remembers."[39] Despite Ferguson's insight regarding Paper X's centrality to the "unfinished" business of publicly representing private meaning in *Clarissa*—an insight carried forward by Macpherson and Lee—scholars continue to interpret this page as a successful representation of psychological breakdown.

This tendency to understand the novel as transparently representing personal trauma has meant that less work has been done to explore the role of materiality in Richardson's "mimesis of distinction." Paper X appears strikingly like a page out of an eighteenth-century commonplace book.[40] What happens when we consider it as an artifact of material culture rather than an artifact of psychological breakdown? Ferguson mentions the "physical" destruction of the printed page in the case of Paper X, but Richardson's decisions here also point to what Deidre Shauna Lynch refers to as an "appreciation of the forces of dispersal and division," of the taking apart and spreading of the books in which the quotations initially appeared.[41] Pace Ferguson, then, agreed upon social "forms" are both symbolic *and* material, such as the woman's body, or the codex form.

We rely on existing legal, political, *and* literary agreements for our conclusions regarding what forms mean and who or what has the ability and credibility to decide their meanings. Shapin and Schaffer argued that because masculine modesty lent credibility to recorded observations it became part of the social form of early scientific fact. Such stipulated social forms can thus determine the shape knowledge takes: what counts as knowable and how one goes about knowing it. Richardson resists preexisting legal and literary-formal agreements because of what they deny in the case of rape. The legal agreements or stipulations discussed by Ferguson do not simply render unimportant the testimony of rape survivors; they render that testimony unreal and nonexistent. Similarly, the stipulation that readers act as if print is handwriting when reading epistolary fiction renders the potential *personal* effects of impersonal reproducible print impossible to see. It is my contention that Richardson turns to literary generalizations in the climactic post-rape Paper X and elsewhere in the novel not only to mark a paradox of interiority's representation (as Ferguson and others have argued), but to reopen debate, making room for disagreement regarding the formal-material compositions of our shared reality, our common world. Generalization is central to this mode of realism as process, as a composition that is changeable, expandable, and inclusive of the "unreal."

Paper X advances this realism of literary generalization in multiple ways. First, it is both personal and impersonal; its origins are ambiguous. Paper X comes from Clarissa, but also from Dorcas who pieces the torn scraps together in the particular order printed, and also from the books that provided Clarissa with these quotations in the first place, books accessible to a reader outside the fictional world. Rather than reveal an interior exposed, these literary generalizations are the bases of a reality that Clarissa shares with her readers—that they compose together.

I am not arguing that maxims bring readers into contact with reality by operating as an instrument of empirical perception. In doing so I would only be asserting what William Warner calls another "discrete and circumscribed literary way to know reality" that was particularly tied to English empiricism and scientific practice.[42] Neither am I claiming that generalized and generalizable forms such as the maxim or a literary quotation taken out of context reflect a premodern connection to communal meaning.[43] As demonstrated in this chapter's first section, novelists and critics were fascinated by the newly constructed unreality of general maxims, even if they primarily summoned this unreality as a contrast to the reality effects of particularized fictional circumstances and experiences. Yet literary generalizations achieve their own distinctive effects in a world of increasingly interiorized truth. Literary generalizations are commonly received but leave no impression on the mind; they are guests that are everywhere and nowhere, opposed to the "Modern" model of knowledge that requires the "bifurcation" of nature and politics for its credibility—a model of knowledge that requires nature to be unassailable by accusations of political (including literary) interference. Instead, literary generalizations advance the possibilities of unpredictable associations through their generality and commonness and thus implicitly raise the question of whether and how other existing associations or "social forms" might be more attuned to the possibilities of alternative accounts of things.

Literary Generalization: Everyone in a Heavy Grief Thinks the Same?

While in Richardson's earlier narratives "the pleasures of imaginative expansion and of punishment and reward run in parallel," as Jacob Sider Jost puts it, this is not what happens in *Clarissa,* a novel that denies the pious heroine not just marriage but life.[44] When she designs the coffin that she will sleep beside in the final weeks of her life at Smith's, Clarissa retroactively gives April 10—the day on which she leaves her father's house with Lovelace—as the date of her death. Her health remains precarious throughout the entirety of the portion of the novel following her return to Smith's. Belford is present for a visit from an apothecary who assures him that Clarissa "would recover if she herself desired to recover, and would use the means" (*C* 1127). As he relates his account of these circumstances to Lovelace, Belford's sentences shift toward sententiousness with the inclusion of italics: the apothecary and doctor "depend too much upon her *youth* . . . and upon *time*" (1127). The tool of medical practitioners is this distilled experience in the form of unstated aphorisms ("Trust to youth" or "Time alleviates suffering"). Belford explicitly contrasts this professional, practical knowledge with literary knowledge: "Her grief, in short, seems to me to be of such a nature, that *time,* which alleviates most other persons' afflictions, will, as the poet says, *give increase to hers*" (1128). The "poet" from whom Belford draws this truth is Congreve. In his Mediterranean tragedy *The Mourning Bride* (1697), the character of Almeria, Princess of Granada, finds herself in the unique position (which is *not* Clarissa's) of grieving the *end* of an incarceration. It is as a prisoner of war that Almeria discovers requited love, both parental and conjugal. Her imprisonment by the King of Valentia was a tender one, and she is desolate upon being carried back to her father's court where she hears that her father-in-law has been executed and that her husband, enemy to her father's nation, is now at large. Her attendant Leonora attempts to calm her: "For Heaven's sake, dear Madam, moderate / Your Griefs, there is no Cause—." Almeria, however, is inconsolable: "Peace—No Cause! yes, there is Eternal Cause, / And Misery Eternal will succeed."[45] When Leonora again wishes for calm on behalf of her mistress ("Look down, good Heav'n, with Pity on her Sorrows, / And grant, that Time may bring her some Relief") Almeria replies: "O no! Time gives Encrease to my Afflictions."[46] We keen modern readers might think that this bit of narrative context should have warned Belford off of turning to Almeria's grief in order to predict how grief might affect Clarissa. Almeria's case is more than nonequivalent to Clarissa's; the conditions of Almeria's sufferings approach the negative image of our heroine's. Almeria has been married; Clarissa has been raped. Clarissa falls into familial persecution, but she encounters her enemy, not her lover, in the enemy of her family. Clarissa's adoption in her meditations of Job's story as analogous to her own is much more apt.

Whence comes this demand for similar conditions that I'm placing on Belford's generalizing use of a literary quotation? It is easy enough to interpret as ornament the inclusion of a literary sentiment by an educated man of Belford's stature. And Belford (and Richardson, too, of course) needs to provide additional support to a conclusion about Clarissa that he knows others will reject. But he finds in Congreve not a turn of phrase or snippet of wit. The poet gives Belford a way of restating what he witnesses of Clarissa's health and of defending the reality of her suffering: the way of the literary. Richardson orchestrates this so that we readers of fiction may watch another reader of fiction deploy a portion of what he's read *against* the words of a doctor, bringing Almeria's claim alongside Clarissa's case, and composing associations as he goes. By the time Belford draws on the words of Congreve's Almeria, he has *seen* the scraps of Clarissa's postrape pages. Lovelace sends them to him, prefacing the copied passages with the following maximic reflection: "Preserve them, therefore. For we often look back with pleasure even upon the heaviest griefs, when the cause of them is removed" (*C* 890). Is Belford following Clarissa's lead of using literary generalizations to create new associations? The apothecary prescribes time, as does Lovelace. But Clarissa's grief is "of such a nature" that it will respond to time differently than most types of grief, growing rather than diminishing. Belford's innovation as he generalizes from this particular literary example is to replace conventional wisdom with an unconventional literary claim. Indeed, he takes this idea from Almeria herself, whose response to Leonora's consolatory generalization that "Time may bring her some Relief" is "Time gives Encrease to my Afflictions."

It is relevant that Lovelace's incredulous response to Belford's use of the quotation appears in a section of the novel in which Belford is introducing Lovelace to Clarissa's extracts from biblical texts:

> *Time*, in the words of Congreve, thou sayst, *will give increase to her afflictions*. But why so? Knowest thou not, that those words (so contrary to common experience) were applied to the case of a person, while passion was in its full vigour?—At such a time, everyone in a heavy grief thinks the same: But as enthusiasts do by Scripture, so dost thou by the poets thou hast read: anything that carries the most distant allusion from *either* to the case in hand, is put down by both for gospel, however incongruous to the general scope of either, and to *that case*. (C 1144)

Here Lovelace emphasizes his understanding of the difference between a claim supported by "common experience" ("Everyone in a heavy grief thinks the same") and a claim articulated in context in writing, the meaning of which depends on the work's "general scope." Both the concept of "common experience" and the concept of "general scope" require one to eliminate outliers and exceptions in order to make the broad category cohere. Lovelace makes the mistake that Diderot makes in his reading of Richardson, assuming that all readers read for one recognizable, credible, universal human experience of the real despite the obvious discrepancies between one person's life and the lives of others (whether real or fictional). Diderot writes of Richardson's fiction, which repeatedly focuses on women's experience of sexual violence: "The passions he depicts are recognizable as the passions I feel myself; they are caused by the same objects and act with a familiar power; the misfortunes and afflictions of his characters are of the same kind as those that constantly threaten me."[47] No, Diderot; no, Lovelace: not everyone in a heavy grief thinks or feels the same.

In Lovelace's move to a generalization ("Everyone in a heavy grief thinks the same"), we ought to be reminded of my earlier example in this chapter of Lady Mary Wortley Montagu's rejection of Clarissa's maxim (a maxim attributed to the heroine but originally formulated by Lovelace).

In order to reject Clarissa's maxim Montagu creates her own: "In this mortal state of imperfection fig-leaves are as necessary for our minds as our bodies, and 'tis as indecent to show all we think, as all we have." In the face of a maxim's call for greater attention to matters of concern applicable to some persons and not others, there is a constant move to countermaximize, to defuse the original generalization with another rather than follow the first to where it leads: on a tour through daily unrealities, to daily denials of reality.

As we have seen in the case of maximizing by Lord M. and Parson Brand, there is no one consistent mode of application of a textual scrap (literary, scriptural, or otherwise) to a "real" or ordinary experience within the world of the novel. Early on Lovelace and Clarissa frequently demonstrate the place of reading in their everyday lives, noting literary quotations "appropriate" to different moments as if to say: *We've seen this in human behavior before.* Clarissa's meditations—timely compositions of well-known and widespread verses, mostly from Job, with the pronoun "her" replacing the pronoun "him"—is of this type. The meditations are a superb example of how such compilations can lend a quality of the extraordinary to the ordinary as represented in the new fictional histories. In one Clarissa writes: "Wherefore is light given to *her* that is in misery; and life unto the bitter in soul?" (*C* 1125). Critics have tended to interpret Clarissa's meditations as they have interpreted Paper X: as acts in which Clarissa speaks through another's lips, and by another's authority, as it were.[48] The meditations are simultaneously anthologies (of scripture) and anthology pieces.[49] Kathryn Steele connects Clarissa's meditations with the earlier "mad letters" on the basis of what she sees as their shared effort to "frustrate communication" and to stand as "material signs of [Clarissa's] withdrawal from the world."[50] The meditations, in particular, are "her private way of communicating with God and of preparing for death" and as such, "deflect or distance the reader from meaning."[51] Belford's application of Congreve's text is also a "material sign," but one of a different type—one designed to expand our shared reality by including within its fold the nonrealities systematically and historically excluded in order to protect the "natural" from charges of ideological construction.[52]

Lovelace accuses literature of functioning as Belford's "Scripture," and, according to Lovelace, Belford is an "enthusiast" who treats one text or set of texts as the interpretive master key for unlocking all meaning. But *The Mourning Bride* is no master key; it is only one among many "Modern" works on which Richardson's various characters draw. *Clarissa* opens a path for thinking of the real as composed and always capable of being recomposed well or poorly. But, aware of its status as literary, and thus as only part of a much larger ongoing composition, *Clarissa* also registers the dangers of embracing an aesthetic that we might call, following Macpherson, a nonmaterialist realism invested in the fantasy of mind's transcendence over matter.[53]

Some scholars have characterized novelistic realism as a descriptive mode linked to empiricist protocols, while others have taken it to be an exercise in realizing subjective experience on the page. *Clarissa*'s realism of literary generalization falls into neither of these categories. The realism of generalization explored here is an assemblage or composition, directing us away from the normative "person" as primary arbiter of reality and refusing to elevate an idealized, autonomous mind as the primary engine for justice. Literary generalizations are common. They are embedded in contexts, and yet they move easily across space and time. Their commonness is linked to a materiality that under empiricism becomes uniquely unreal, because it is held as incompatible with both interiority and with nature purified of ideology. As knowledge, these generalizations become the property of no mind; this is an asset, because no mind is independent of matter.

Literary generalizations in *Clarissa* make available accounts of the "real" that are collaborative, and thus potentially more progressive. Yet a materialism or reality that is "new" is not necessarily more just.[54] What we should appreciate in *Clarissa*'s realism of literary generalization is the invitation it extends to us—its readers and fellow makers—to see ourselves as public actors with very real and often competing interests and agencies that help constitute the material conditions under which we flourish or suffer.

4
Austen's Lessons Not Worth Knowing

"I WISH YOU COULD DANCE, my dear,—I wish you could get a partner"; "I wish we had a large acquaintance here"; "I wish I had a large acquaintance here with all my heart, and then I should get you a partner"; "how pleasant it would be if we had any acquaintance here."[1] This refrain is uttered by Catherine Morland's companion, Mrs. Allen, in their first days at Bath. We readers are eager to follow Catherine, heroine of *Northanger Abbey* (1817), through "all the difficulties and dangers of a six weeks' residence" in that city—a period the narrator frames as the young woman's "entrée into life" (8, 10). Prior to Catherine's arrival, Austen's narrator has reminded readers of the events that occur in gothic versions of such stories: the violent "noblemen and baronets [that] delight in forcing young ladies away to some remote farm-house," the "robbers," "tempests," and "lucky overturn[s] to introduce them to the hero" (9, 10). Instead, Catherine's first difficulty is only the irritation and discomfort—no less felt for being banal—of half-belonging among an unknown public, of "looking at every body and speaking to no one" (14).

Mrs. Allen's repeated wish for acquaintance receives extended narrational treatment and is ultimately rewarded by the author who pulls the strings of this fictional world:

> This sentiment had been uttered so often in vain, that Mrs. Allen had no particular reason to hope it would be followed with more advantage now; but we are told to "despair of nothing we would attain," as "unwearied diligence our point would gain"; and the unwearied diligence with which she had every day wished for the same thing was at length to have its just reward, for hardly had she been seated ten minutes before a lady of about her own age, who was sitting by her, and had been looking at her attentively for several minutes, addressed her with great complaisance in these words:—"I think, madam, I cannot be mistaken; it is a long time since I had the pleasure of seeing you, but is not your name Allen?" (18–19)

The passage contains that now-familiar eighteenth-century structure: a particularized example paired with a maxim. Example and maxim: two different modes for rendering truth, side by side. With that phrase "unwearied diligence," which appears first in the quoted maxim and then in the narrative report of Mrs. Allen's actions, Austen cinches maxim and example together while insisting on the seam. The maxim states a principle we wish were true: try hard enough and you will be rewarded. Its presence taunts us here, because although Mrs. Allen has demonstrated no real diligence to deserve it, she nonetheless receives her "just reward."

Because Mrs. Allen so clearly contrasts with our heroine, we might be tempted to interpret her as a foil—a flat character sketch of an ignorant woman. The proximity of the maxim to her pointless speech encourages this interpretation. As previous editors have noted, this maxim originated in Thomas Dyche's *A Guide to the English Tongue* (1707), an "oft-reprinted educational book . . . designed to teach children proper pronunciation and spelling, via moral verses to be copied out and recited."[2] At least one scholar has assigned this maxim to the mind of Mrs. Allen, presumably as an unusual example of free indirect thought. Austen could not be claiming such drab reading material as her own.[3] Indeed, when we consider another contrasting duo—the morally earnest Mary Bennet alongside her witty sister Elizabeth—Austen seems to suggest that maxims are the refuge of small, uninteresting minds, especially the minds of certain women. We need think only of Mary's pointed pronouncements on decorum, such as her statement that "every impulse of feeling should be guided by reason; and, in my opinion, exertion should always be in proportion to what is required," to remember that direct didacticism fails to deliver that distinctly Austen thrill of suggestive creative possibility.[4] Elizabeth Bennet and even Catherine Morland, by contrast, are unafraid of the searching journey of private self-reflection and need no mental space-filler.

According to this reading, the above passage's maxim is simply an extension of pointless speech. It is the transcription of unperceptive, unimportant thought—the kind that *could* belong to Mrs. Allen, but does not necessarily belong to her specifically.[5] Many of Austen's readers would have recognized this maxim from their school days and could have completed its rhyme from memory. When Austen uses the "but we are told" to introduce it, she hails this wider audience composed of educated middle-class women. Because of the sentence's passive construction, there is no mention of who does the telling. But the "we" is a community of listeners, readers, and rememberers to which the author-narrator belongs. It is unlikely that Austen means for us to believe that this maxim rises to the front of her character's mind, drawn from remembered reading. Mrs. Allen is not repeating herself because she was taught to do so. The passage's joke arises from the maxim's ironic placement. Despite Mrs. Allen's unprincipled action, the maxim's placement assigns her agency: "The unwearied diligence with which [Mrs. Allen] had every day wished for the same thing was at length to have its just reward." It is not the maxim, but the author who rewards Mrs. Allen, and who does so for no good or just reason, but because she desires it. That is the power this author has and—importantly—she flaunts it with language that reminds us of her schoolgirl past.

Austen—that wielder of "Absolute Style," the "cool, compressed adequation of language to whatever it wants to say"—is willing to trade in a form that, because it is general, seems imprecise with regard to individual experience, even perhaps to her own.[6] And yet, as I've suggested, moral maxims of the sort that intersect with Mrs. Allen's empty mind belong to the reading experience of this author. Austen criticism has long been preoccupied with the way in which her plots and style smooth over tensions between individual and society, or, more poignantly, between the secrets of a woman's ambitious heart and the public facts of her life. Given the pressing need for such reconciliation for readers, why would Austen, with the maxim, choose to remind them that the institutions that early fostered their intellects provided them with such indirect and insignificant modes of expression?[7] By the time Austen was writing, moral maxims and extracts were considered insufficient educational tools that encouraged superficial intellectual development. In *Strictures on the Modern System of Female Education* (1799), for example, Hannah More attacks "the hackney'd quotations of certain accomplished young ladies," ladies that had begun modeling their reading off of those procedures promoted by didactic novels themselves.[8] Austen nods to this critique and others like it when, at the beginning of *Northanger Abbey,* her narrator dutifully reproduces in commonplace form the "quotations" Catherine has extracted from literature and stored for use in her life as a heroine (7). As Leah Price points out, More's "concern [was] epistemological as much as moral": "Quoting reflects not simply feminine [intellectual] vanity, but feminine imposture."[9] The mind full of maxims was not full of knowledge, and to teach women in this way was to teach them to *pretend* to know without the actual mental effort of extensive study.

By the time Austen was drafting *Northanger Abbey* in the 1790s, both the culture of the maxim *and* its underside—the subculture of the self-canceling maxim—had come a long way. In the late seventeenth and early eighteenth centuries, Locke kept manuscript commonplace books, as did plenty of educated women. A person could cull principles from political treatises and moral truisms from printed sermons and oft-printed works of personal devotion. They could copy quotes from translated French romances and poetic miscellanies. They could even purchase prearranged print commonplace books.[10] Many of these early devotional works and collections of practical morality were designed for busy, educated, working men, and counterpart conduct books existed for their wives, concerning the woman's role in the rising modern domestic sphere.[11] Such are the books from which Richardson's devout Pamela, that harbinger of the economic and social advantages of middle-class feminine English virtue, quotes.[12] The first two chapters of *Maxims and the Mind* aspired to send such religious and conduct-book maxims to the background by highlighting instead early perversions of the form: maxims as vehicles of radical philosophical discovery (Bacon's aphoristic knowledge broken) and feminine creative sexual display (Behn's ventriloquy of Amynta-La Rochefoucauld). I outlined the philosophical and gendered stakes of such perverse departures from standard rule, and in tracing them to Swift's and Richardson's fictional responses, I demonstrated their relevance to early novelistic inwardness. Swift and Richardson creatively shaped alternative, maximic spaces with unclear boundaries between interiors and exteriors in response to insistent pressures of embodiment and socially sanctioned meaning. Swift's Tubman in a garret needs medicine, needs to eat. After he rapes an unconscious Clarissa, Lovelace insists that her body may still manifest signs of her consent. Yet how can a novel realize its heroine's mental state of nonconsent when rape law has already eradicated her autonomous personhood and thus her ability to freely consent? What happens to people who must read and write to survive, who rely on expression for their very existence? What opportunities for escape do they have? I have argued that some writers, faced with such pressures, made minds out of maxims, and in doing so presented fictional mental interiors that were disordered collections of unowned ideas *instead* of a penetrable space of individual identity, the seat of personal autonomy.

I have sought, in other words, to make visible some alternatives to Nancy Armstrong's story of the relationship between novelistic consciousness and the socially marginalized writer—particularly, for Armstrong, the woman who writes. As I move into an examination of Austen, however, it is necessary to sketch a bit more of the cultural shifts in reading's role in education over the course of the eighteenth century. Leah Price's work on the role of the anthology in educational practices at this time is particularly helpful, in that Price outlines the arguments made by late eighteenth-century anthologists and compilers for teaching through parts (e.g., maxims or poetic extracts), not wholes (treatises).[13] Vicesimus Knox's *Elegant Extracts: Or Useful and Entertaining Passages in Prose Selected for the Improvement of Scholars in Classical and Other Schools* (1784) and his companion volume, *Elegant Extracts: Or Useful and Entertaining Pieces of Poetry, Selected for the Improvement of Youth in Speaking, Reading, Thinking, Composing; and in the Conduct of Life* (1784) took advantage of the end of perpetual copyright in 1774 to pitch the literary anthology not only as a canon-making mechanism but as a technology for capturing public opinion and taste.[14] According to Knox, the anthology spoke the voice of "public" rather than "private" judgment.[15] Knox, himself classically educated, leaned into the cultural shift away from pedagogical approaches differentiated by gender—with boys learning the ancients and girls the moderns—to one in which youths of both sexes could use the same educational material. Furthermore, Knox replaced references to a "common" reader with allusions to a distinctly middle-class commercial public. What Hannah More reacted against by the end of the eighteenth century, according to Price, was "the convergence that anthologies like Knox's . . . celebrated between the feminization of schooling and the commercialization of literature."[16] Austen was clearly interested in such shifts and debates, and in *Northanger Abbey* she considers them from the perspective of individual, socially situated readers.

Austen, this chapter argues, found purpose in the shared memory of silly moral maxims and extracts. With such maxims, her narrator records the exact words that thousands have read or heard and maybe even entertained—some sincerely and some with a heavy dose of skepticism. In other words, a maxim can be, at one and the same time, a transcription of speech heard or words read, *and* a transcription of a thought that rises as memory for an individual in response to events in her world. We remember a maxim and it triggers other memories of where, how, and why we read it before. As children we might have learned to recite "Despair of Nothing that you wou'd attain: / Unweari'd Diligence your Point will gain," and, though proud of our accomplishments, we might not have really understood what the words meant. As adults we encounter it again, remember its form and recognize—perhaps for the first time—content we previously missed. The maxim may inspire a new application to our life, linking past ignorance to present uncertainty. To the extent that the event and articulation of a remembered maxim is unique to each person, it is unique only for the time and context in which it arises—and, indeed, each context will always be different for everyone.

Austen's maxims—from Mrs. Allen's to Mary Bennet's to Elizabeth's— capture shared acceptance of "good enough" compact language through a form that, because it is common and widely recognized, is spacious enough to be adequate for capturing (in context) many different experiences of which it can speak. Austen recognized that although maxims are general touchstones, they are also linguistically precise. Readers will know from their own experience these exact words in this order, albeit placed in a context unique to their lives. Put another way, Austen's maxims gesture at a depth of private, individual experience *and* at a poignancy beyond it— the poignancy of getting "close enough" to an articulation of one's authentic experience that someone else can understand. Because the maxims are often dully didactic, drawing a connection between education and moral worth, Austen's playfulness with them (she takes them seriously by making fun of them) stretches to encompass an interest in ignorance and how ignorance functions socially for women.

The Conduct Book, the "Consciousness Scene," and the Maxim

As a novelist ever conscious of her eighteenth-century forebears, Austen recognized the centrality of maxims to book learning and their simultaneous, uneasy presence within early novels (such as in the novels of Richardson). Richardson, initially disclaiming original authorship but not the moral authority of his works, was drawn to maxims.[17] Yet the maxim's impersonality and alignment with rhetorical deceit led him to avoid identifying it as his paragon's primary moral tool. As a similar creator of independent, ethical women characters, Austen shares Richardson's concerns about the maxim's rhetorical function, and she distinguishes between the moral givenness of a maxim and the nuances of judgment required of women navigating the social terrain of the early nineteenth-century gentry.[18]

In the early nineteenth century, writers of realist fiction remained pressured to address education directly, particularly the education of women readers. This concern with education fueled the ongoing production of didactic fiction in which maxims remained tools for delivering instruction regarding proper feminine conduct. In scholarship on late seventeenth-century, eighteenth-century, and early nineteenth-century women's writing, maxims have been strongly associated with the conduct book's mode of moral education. Many, but not all, conduct books did draw explicitly on the form of the moral precept. George Savile, Marquis of Halifax's *The Lady's New-Year's-Gift, or, Advice to a Daughter* (1688) is one example. Thomas Marriott's 1759 *Female Conduct* is another. When Mary Bennet, the most obvious of Austen's maximizers, delivers a maxim, she does so in the explicit tradition of Savile or, later, Fordyce, seeming to speak directly from their books. Compare, for example, Mary's commentary on pride—that "vanity and pride are different things, though the words are often used synonymously.... Pride relates more to our opinion of ourselves, vanity to what we would have others think of us" (21)—to Savile's statement in the late seventeenth-century conduct book, *The Lady's New-Year's-Gift*: "After having said this against *Vanity*, I do not intend to apply the same *Censure* to *Pride*, well placed, and rightly defined. It is an *ambiguous Word*; one kind of it is as much a *Vertue*, as the other is a *Vice*."[19] Such lessons encouraged consideration of moral behavior through the study of abstract concepts in sentences that could be extracted and recorded in one's commonplace book or mind.

Austen's fiction does not convey its moral and social insights through extractable maxims or abstract nouns (despite what titles like *Pride and Prejudice* often lead readers to believe). In her work, character, scene, and narrative interact to undermine the codes and conventions that appeared in eighteenth-century women's conduct books, especially those written for women by men. By strategically imitating and undermining such lessons on propriety and the feminine ideal, Austen demonstrates the absurdity of constraints placed on women of the English gentry and establishes the value of an individual's right to self-determine through acts of independent rational judgment.[20] When Audrey Bilger, for example, examines Austen's radical feminist humor, she focuses on scenes that call attention to categories of behavior touched on in conduct books. Calling our attention to the drawing-room scene at Netherfield Hall, in which Elizabeth and Caroline Bingley "take a turn" around the room, Bilger contends that Austen explicitly has in mind conduct book restrictions on feminine behavior when she has Elizabeth suggest to Caroline Bingley that they tease back, going against "the conduct-book rule that women who laugh at men will 'expose' only themselves."[21] Ultimately, according to Bilger, Austen's subtle critique of the proscription of laughter lends depth to Elizabeth *and* Darcy: "Elizabeth gets rewarded for behaving in contradiction to the conduct-book regulations, and Darcy reveals his sterling character by admiring her strengths. She violates the conventions that guide gender relations, but her ability to do so frees her for the rewards of true partnership—one that will, no doubt, involve its share of laughter."[22] This theory of conduct critique holds that in using particularized and dramatized scenes to reproduce then undermine general rules of propriety drawn from books, Austen renders her characters "deep" because she portrays them as individuals liberated from traditional restraints. The nuanced, indirect critic of conduct posits new ways of being without too directly attacking the system and thereby threatening her reputation.[23]

Yet, despite the maxim's association with conduct books, maxims in Austen's two works most directly parodic of the conduct genre—*Northanger Abbey* and *Pride and Prejudice*—are not obviously and simplistically folded into her parodies. The remainder of this section will look at conduct books in light of the development of the "consciousness scene" in the early novel and consider how Austen's maxims turn us away from the novel's emphasis on feminine self-examination and toward other expressive possibilities.

Paul Dawson demonstrates that the eighteenth-century conduct book's instructions directed at women shaped scenes of consciousness in the modern novel. "The technology of the self encouraged by conduct books," he argues, "provides the cultural impetus and structural frame for the formal method of rendering characters' interiority, establishing the generic scene of heroines reviewing their conduct as a convention that persists into the twentieth century."[24] Conduct books instructed women in the practice of examining the degree to which their external behavior corresponded to a feminine ideal they were meant to internalize. Early novelistic "consciousness scenes" were then informed by this practice. Consider, for example, the language used to display Elizabeth's self-assessment after she rejects Darcy then reads his letter. Initially, Austen relies on psychonarration—"[Elizabeth] grew absolutely ashamed of herself.—Of neither Darcy nor Wickham could she think, without feeling that she had been blind, partial, prejudiced, absurd" (201)—and soliloquy: "'How despicably have I acted!' she cried. '—I, who have prided myself on my discernment!'" (201–22). We see a similar consciousness scene in *Northanger Abbey*, in which reported thought shifts into free indirect discourse when Catherine Morland reflects on her conduct after revealing to Henry her suspicions about his father:

> The visions of romance were over. Catherine was completely awakened. Henry's address, short as it had been, had more thoroughly opened her eyes to the extravagance of her late fancies than all their several disappointments had done. Most grievously was she humbled. Most bitterly did she cry. It was not only with herself that she was sunk—but with Henry. Her folly, which now seemed even criminal, was all exposed to him, and he must despise her for ever. The liberty which her imagination had dared to take with the character of his father, could he ever forgive it? The absurdity of her curiosity and her fears, could they ever be forgotten? She hated herself more than she could express. He had—she thought he had, once or twice before this fatal morning, shewn something like affection for her.—But now—in short, she made herself as miserable as possible for about half an hour, went down when the clock struck five, with a broken heart, and could scarcely give intelligible answer to Eleanor's inquiry, if she was well. (146)

Catherine has been taught her lesson—a grievous one indeed. Dawson's argument is persuasive, as we can see the way in which both Catherine's and Elizabeth's reflections (not to mention those of many other novelistic heroines) occur in isolation—alone outside or in a bedroom—and with their conduct as the object. This time spent in reflection dramatizes the internalization of the process of self-surveillance, such that examination leads to transformation of self. Elizabeth Bennet's persistence in reflecting on her conduct toward Darcy across time, for example, becomes the edifice upon which her narrative of gradual self-discovery is built. This self-discovery is ultimately a process of growth that reveals a "gratitude" that primes her for marriage and simultaneously prepares her for that conjunction of individual and society that will manifest internally (as a self-conception revised according to Darcy's perception of her) and externally (as a member of a heterosexual couple). Dawson writes that "it is precisely this struggle of female characters to reconcile their private desires with expectations of their external conduct that necessitated a rendering of their internal scrutiny, and hence the gradual merger of psycho-narration and the soliloquy in the grammar of third-person narration" (169). To identify errors in conduct is to reveal previously unknown dimensions of oneself. Where ignorance once dwelled, now there is knowledge.

Catherine's consciousness scene, however, is unique, and Dawson does not consider it. The consciousness scenes he reviews arise when solitary women wonder what others must think of their actions. Catherine, by contrast, examines her conduct in the full light, as it were, of knowledge that Henry knows much of what she thought about his father. The interrogatives marking the presence of the free indirect style in this passage from *Northanger Abbey* ask not whether Catherine should have acted differently, but whether Henry can forgive and forget what has happened. The passage as a whole is less about the consolidation of a social self suitable for the married state and more about reckoning, however briefly, with the facets of our selves always on display for those who know us best. Henry *may* never forget the "absurdity" of Catherine's curiosity and fear, but he will forgive her and live with them. Catherine may be miserable, but only for thirty minutes.

We might pause to note here that *Pride and Prejudice* also gives us Mary Bennet, a woman who both has private desires and engages in questionable public conduct, but who receives no nuanced narrational scrutiny and no consciousness scene. It is easy to be dismissive of Mary, and Austen certainly ridicules her. Maximizing gives Mary an air of moralistic self-importance not unlike Clarissa's, but, instead of possessing Clarissa's wealth and beauty, Mary comes from a gentleman's family of moderate income and is plain. Austen caricatures each of the lesser Bennet sisters, but only Mary's talk predictably produces a palpable nothingness. We know that Mary studies and is desirous of public praise, but the only products of her study we see are maxims she delivers in dialogue. Consider Mary's role in an early scene in the novel, when Elizabeth declares her intention to walk to Netherfield Park to visit an ill Jane:

> "How can you be so silly," cried her mother, "as to think of [walking], in all this dirt! You will not be fit to be seen when you get there."

"I shall be very fit to see Jane—which is all I want."

"Is this a hint to me, Lizzy," said her father, "to send for the horses?"

"No, indeed. I do not wish to avoid the walk. The distance is nothing, when one has a motive; only three miles. I shall be back by dinner."

"I admire the activity of your benevolence," observed Mary, "but every impulse of feeling should be guided by reason; and, in my opinion, exertion should always be in proportion to what is required."

"We will go as far as Meryton with you," said Catherine and Lydia.— Elizabeth accepted their company, and the three young ladies set off together. (32–33)

I have included a longer portion of the passage to demonstrate the degree to which Austen situates Mary as a conversational nonentity. While each of the other speakers addresses someone directly, Mary's comment—ostensibly directed at Elizabeth—is delivered to the whole room, to everyone and no one. Perhaps fittingly, no one responds. For all the talk of Austen as a writer of balance, of moral moderation or the Aristotelian mean, Mary's paean to proportion is ironically excessive.[25]

Mary seems to want to prompt internal reflection in all, but her suggestions fall flat amid lively conversation. She fails to grasp the value of private self-surveillance, a secret internal process that makes proper conduct look effortless. Mary does *not* hide her (dim) light in a bushel. Eager for display, she seeks out praise. At the Netherfield Ball, when Mary "after very little entreaty, [prepares] to oblige the company," Jane is busy with Bingley and Lydia and Kitty are distracted elsewhere. Elizabeth's "painful sensations" are, however, so great at listening to her sister's subpar performance—"her voice was weak, and her manner affected.—Elizabeth was in agonies"—that she appeals to her father to urge Mary to stop, which he does in front of everyone. The narrator gives us an externalized glimpse of Mary's feelings at this moment: "Mary, though pretending not to hear, was somewhat disconcerted; and Elizabeth sorry for her, and sorry for her father's speech, was afraid her anxiety had done no good" (*P* 98). The publicity of Mary's conduct lessons creates the illusion that she, unlike Elizabeth and Catherine, lacks self-insight. And yet there is no hard evidence of this in the novel, and Mary is certainly alone in her room enough to justify moments in which she would evaluate her conduct and her self. Unlike Lydia and Kitty, Mary seems to feel her public mistakes; unlike Jane, she makes them. The novel's narrative perspective hews close to Elizabeth's consciousness, and Austen simply gives no other woman a consciousness scene.

We first meet Mary shortly after the famous opening line to *Pride and Prejudice*, as a minor character within the novel's initial scene. The received wisdom about single men that Austen ironically delivers in that first line leads directly into a scene in which a man displays his social advantage relative to women by making them feel dumb. Mrs. Bennet is eager for her family to make the acquaintance of Mr. Bingley, and the rules of conduct dictate that her husband must pay the initial visit. Although Mr. Bennet plans to visit, he tells his wife and daughters that he will not. After he makes Bingley's acquaintance, he then teases the women for not knowing that he did. When Mrs. Bennet complains that she will have to meet Bingley publicly through a friend who has met him first, Mr. Bennet informs his wife that she must introduce her friend. "Nonsense, nonsense!" cries Mrs. Bennet. The following passage relates her husband's response:

> "What can be the meaning of that emphatic exclamation?" cried he. "Do you consider the forms of introduction, and the stress that is laid on them, as nonsense? I cannot quite agree with you *there*. What say you, Mary? for you are a young lady of deep reflection I know, and read great books, and make extracts."
>
> Mary wished to say something very sensible, but knew not how.
>
> "While Mary is adjusting her ideas, let us return to Mr. Bingley." (9)

It is a funny scene, engaging for the reader who does not share the women's confusion. Indeed, before the above scene takes place, the Austenian narrator matter-of-factly delivers this information to the reader: "Mr. Bennet was among the earliest of those who waited on Mr. Bingley. He had always intended to visit him, though to the last always assuring his wife that he should not go; and till the evening after the visit was paid, she had no knowledge of it. It was then disclosed in the following manner" (8). It is notable that Mr. Bennet specifically jokes about the "forms of introduction," which he funnels through a second joke at Mary's expense. Mary, of course, cannot say anything "sensible" in response, because the "extracts" from her "great books" are too general to help her navigate this level of sarcastic obfuscation. In this context, her knowledge renders her ignorant, a bind that Mr. Bennet takes advantage of by virtue of his social position.

The Benefits of Ignorance

Pride and Prejudice figures Mary as an outsider. When she speaks, no one responds, and, unlike the Lydia-Kitty and Elizabeth-Jane pairings, she is uncoupled, an isolated sister, not particularly loved by either parent. Yet, despite this social status as outsider, the style of discourse most frequently ascribed to Mary renders her something of an everywoman. She speaks the language of general truth; she is a woman of maxims. What she says might not be wrong, but it doesn't *land*. She does not provide the kind of vicarious experience that we contemporary readers assume all readers crave.

The consciousness scene and novelistic representations of interiority in general are oddly positioned on the ignorance to knowledge spectrum. On the one hand, the novel's representation of invisible thought is considered one pinnacle of the genre's realism. On the other hand, theorists of free indirect discourse, the most developed method of thought representation, portray it as representing something blurry and unclear. Dorrit Cohn argues that the technique of free indirect thought (a subcategory of free indirect discourse) does not reveal a mind but rather, creates the effect of viewing a partially obscured one: "Narrative language appears as a kind of mask, from behind which sounds the voice of a figurative mind."[26] There is also the possibility—explored by Anne-Lise François—that third-person narration relates thoughts about which the character is not aware.[27] Indeed, François argues that third-person narration and free indirect discourse, in the work of Austen in particular, "have special pertinence to the problem of thoughts and wishes that cannot withstand the work of articulation, because they leave in question the protagonist's relationship to the thought and speech acts attributed to her and assume no necessary connection between stated and lived experience."[28] What I am suggesting with relation to Mary is that her speech acts, in that they speak generally to *everyone*'s experience and not to hers in particular, could also be interpreted in this way—as leaving in question her relationship to the thoughts attributed to her such that, beyond her having read the thing she repeats, there may be no other "necessary connection between stated and lived experience." We might say that Mary *knows* what she says but is always at least partially ignorant of what it means for her own experience. Austen is intrigued by the combination of knowledge and ignorance conveyed by this type of speech, because she returns to it repeatedly and alongside the broader cultural issue of feminine ignorance.

Mary strives to be seen by others as learned, and thus she does not benefit from ignorance in the ways another Austen heroine does. When we first meet Catherine Morland of *Northanger Abbey*, we are told that she is in need of direct, even forced, instruction.²⁹ The narrator pulls no punches. Catherine is a good girl who is occasionally dense: "[Catherine] never could learn or understand any thing before she was taught," we are told, "and sometimes not even then, for she was often inattentive, and occasionally stupid" (5). No autodidact, Catherine moves grudgingly through her childhood lessons, and "sometimes" not even they make a dent. She detests "The Beggar's Petition," an oft-memorized schoolhouse poem that begins, "Pity the sorrows of a poor old man!" whose despair originates in a daughter's disgrace. Her "mind [was] about as ignorant and uninformed as the female mind at seventeen usually is" (9). When Catherine arrives in Bath she discovers even more evidence of her unfamiliarity with further subcategories of learning. When Henry and Eleanor Tilney discuss the Bath landscape using the terms of the picturesque, Catherine feels "heartily ashamed of her ignorance" (81). She is embarrassed, believing that her inability to admire natural beauty according to established aesthetic rules has revealed her social inferiority. Henry, however, seems charmed by Catherine's authenticity and innocence. She does not—because she cannot—hide her actual judgments behind the terminology of expertise.

In case we have failed to draw these conclusions about Henry's view of Catherine, Austen's narrator immediately steps in to say that Catherine's ignorance is a romantic advantage. Her shame is "misplaced," because "where people wish to attach, they should always be ignorant." Of course, the context of the narrator's commentary casts doubt onto the proclamation, for this narrator is a self-identified novel-writer who has already declared to her many women readers that novels convey "the most thorough knowledge of human nature, the happiest delineations of its varieties ... in the best chosen language" (*NA* 24). The novel-writing narrator, that is, tells her readers that ignorance is attractive to men at the same time that she delivers knowledge to them. Although she feels stupid (and says so), Catherine has an advantage over other women who possess more learning than she does but who have to hide it: "To come with a well-informed mind, is to come with an inability of administering to the vanity of others, which a sensible person would always wish to avoid. A woman especially, if she have the misfortune of knowing anything, should conceal it as well as she can" (81). This quotation stands as one of the strongest examples of sharp irony in Austen and is typically paired with a similar comment about herself drawn from Austen's correspondence: "I think I may boast myself to be, with all possible Vanity, the most unlearned, & uninformed Female who ever dared to be an Authoress."[30] Middle-class women like Austen were advised to be well informed and then hide it—to know but to seem not to know.

Within the history of the early novel, female inexperience was frequently and repeatedly rendered narratively productive. The entrance of an inexperienced character into the world enables a writer to demonstrate minutely the effects of experience on a relatively untouched mind. In a typical novel of manners, the heroine's inexperience leads to mistakes that become opportunities for learning. We see this coming-of-age trajectory in *Northanger Abbey*. Catherine Morland's behavior is on display for a reader's judgment and, ultimately, edification. Yet Catherine's gradual instruction in worldly social behavior is not the point. Rather, recognizing the quality of her "ignorance" is. Catherine is unsure about codes of politeness. When attempting to apologize to Miss Tilney for breaking their engagement, Catherine is told the latter is not at home. When Catherine then sees Eleanor issue from her Bath residence, she "remembered her own ignorance": "She knew not how such an offence as her's might be classed by the laws of worldly politeness, to what a degree of unforgivingness it might with propriety lead, nor to what rigours of rudeness in return it might justly make her amenable" (66). Ultimately, *Northanger Abbey* does not align social inexperience with moral ignorance, and, rather than coming to disdain the hypocrisies of politeness, Catherine simply learns from Eleanor Tilney that the latter was at the mercy of her father: General Tilney was eager to leave and instructed the servant to declare his daughter not at home. Catherine's moral strength—her sense of the importance of keeping promises—comes from within. When, for example, John Thorpe tries to make "safe" Catherine's "conscience" by "ma[king] [her] excuses" to the Tilneys without Catherine's knowledge, Catherine declares: "I cannot submit to this," and "if I could not be persuaded into doing what I thought wrong, I never will be tricked into it" (72, 73). Thanks to her internal moral compass, Catherine gradually learns that her trusted social guides are often wrong or simply tell untruths to serve their own interests.

In quixotic novels, a novelistic subcategory in which *Northanger Abbey* is often placed, inexperience can take the shape of misperceptions of the world based on a lack of experience coupled with impressions caused by specific types of reading. The simplest reading of *Northanger Abbey* finds that Catherine's quixotic tendencies come from her reading of gothic fiction. She believes General Tilney is a wife-murderer even though the truth is that he was not the best husband. Her discovery of her own misperceptions is part of her personal movement from inexperience to experience and from adolescence into adulthood. Recently, scholars have troubled these assumptions regarding feminine inexperience and the structure of the bildungsroman, examining forms of otherworldly feminine unknowing that strengthen moral resolve. In such cases, ignorance of worldly things enables a counterfactual approach to the world that opens visions of possible, more just futures.[31] Approaching the issue of eighteenth-century heroines' apparent misperception and misjudgment from another angle, Aaron Hanlon has argued that such perceived feminine errors are not errors at all. Hanlon points out that an "(ironically) quixotic move" is common in interpretations of quixotic novels, "which is to collapse [the heroine's] experiences in the world into her experiences of reading."[32] This mistake prevents critics from seeing that the "data" that influence the behavior of a female quixote such as Arabella from Charlotte Lennox's *The Female Quixote* (1752) is "both textual and experiential"—drawn, that is, from books *and* from life. The "complex interaction between [these] two forms of data" is one reason why eighteenth-century authors found the quixotic character so useful for interrogating realist fiction's empirical mechanisms. It makes more sense, Hanlon argues, to consider quixotic novels as presenting us with two "data sources," life experiences and knowledge from books, the interaction of which can explain a character's behavior: "Taken simply as satire, the quixotic motif in fiction is fundamentally about obsessive or delusional over-reading of anachronistic or aberrant source material, but the effect of that satirical gesture is produced by a distinct mechanism, the mismatch between what the quixote perceives and what everyone else perceives. To explain even the conventional reading of quixotism as satire, then, requires an explanation of that empirical mechanism on which it rests."[33] For Hanlon, novels can function epistemologically in their own right, offering "a novelistic interrogation of the reliability of sense perception as a basis for knowing."[34] In *The Female Quixote*, for example, Lennox does not theorize the empiricist problem of perception "as did Locke, Berkeley, and Hume," but rather uses quixotism "to illustrate and draw out the problem, to push empiricism to its logical limits."³⁵ Arabella's perceptions are shaped and limited by the social and legal conditions under which her life as a woman is lived. It is not some innate weakness in her female faculties that leads her astray or makes her more susceptible to the quixotic worldview. Many of the conclusions she draws have some basis in her own reality, and she reinforces her worldview by repeated sensory observations that support—or seem to support—it.

While Hanlon's approach to empirical knowledge in the eighteenth-century novel usefully expands the topic beyond realist representation, Austen's *Northanger Abbey* does not follow the same pattern of critique launched by Lennox's novel, despite its shared interest in the female quixote. The empirical data on which Catherine Morland bases her judgment of General Tilney's evils—his irritability, his domineering treatment of his son and daughter, the weight of silence bearing down upon all parties in his presence—simply do not support her inferences about his homicidal tendencies. Repeatedly, especially in her early fiction, such as the juvenilia and *Northanger Abbey*, Austen's narrator draws us away from the particularities of sensory perception by shifting into an imitation of a distinctive rhetorical register suggestive of a particular ideological perspective. Such moments frequently occur (as in Swift) at the level of generalization. "Every young lady," the narrator tells us, as Catherine attempts to ditch John Thorpe at a ball, "may feel for my heroine in this critical moment, for every young lady has at some time or other known the same agitation. All have been, or at least all have believed themselves to be, in danger from the pursuit of some one whom they wished to avoid" (52). In line with other gothic bromides in the novel, this one also seems designed to deflate the drama of Catherine's "agitation" by implicitly comparing it to that of a heroine from a novel by Radcliffe or, worse, Lewis, attempting escape from physical, even sexual, violence. At the same time, the subject matter of these generalizations—the universal feeling of wanting to avoid someone—is trivial. This is not to say that Catherine's gothic immersion is meant to have no effect on the reader's view of her "ordinary" English reality. "From the age of fifteen on," as Claudia Johnson writes, "Austen, sceptical and unawed, refuses to be lulled by her medium and is determined to illuminate the interests served by its broadest structural outlines down to the subtlest details of its words, rhythms, and cadences."³⁶ With that nod to a gothic bromide, Austen has shifted us into a politicized genre and aligned that genre with "every young lady's" feelings. The gothic taught, of course, that sensitive young girls could be disobedient and were often right to be so when confronted by tyrannical "protectors."[37] There are, in other words, persons waiting to indoctrinate us into customs that serve their interests, and we women can disobey by refusing to learn. Ignorance is not just a romantic advantage, but a political one.

Austen's Maxims of Love

When the Bennets catch wind of Bingley's return to Netherfield, Jane insists that she is unperturbed. Elizabeth does not believe her sister, but neither does she believe that her sister is being intentionally deceptive. Even the narrator assures us that Jane does not know her own desires: she "declared" what she "really believed to be her feelings, in the expectation of his arrival" (*P* 314). After having the opportunity to spend an evening with him, Jane's spirits rise further, but she still insists on her indifference. In their subsequent private conversation, Elizabeth plays the role of the interpreter of her sister's inner, unspoken feelings. She "suspects" Jane, while Jane, in turn, insists that she no longer suspects that Bingley ever "had any design of engaging [her] affection." Elizabeth continues to provoke her:

> "You are very cruel," said [Elizabeth], "you will not let me smile, and are provoking me to it every moment."
> "How hard it is in some cases to be believed!"
> "And how impossible in others!"
> "But why should you wish to persuade me that I feel more than I acknowledge?"
> "That is a question which I hardly know how to answer. We all love to instruct, though we can teach only what is not worth knowing. Forgive me; and if you persist in indifference, do not make *me* your confidante." (324)

This chapter has been structured around excerpts—the parts of the wholes that command our attention—as so many essays and articles about books are. And because this has been a book about the most extractable of extractable bits, this passage too holds a part within a part: Elizabeth's apologetic quip "We all love to instruct, though we can teach only what is not worth knowing." Aside from this maxim, this is a passage about love—how easy it can be to identify it in another, and how difficult to see it in ourselves. Throughout the novel Jane and Elizabeth have practiced examining their feelings in confidence with one another. Given her disappointment of months ago, Jane knows that opinion and convention are against her. She must love Bingley still. As her response to her sister reveals, however, she resists this interpretation, with a disarmingly honest question: Why should you try to persuade me that I feel something that I say I don't feel? Elizabeth dodges the "you" and defers to the general "we." Perhaps she does so because she is in the midst of trying to ascertain her own "true" feelings about Darcy and Darcy's feelings about her. Austen does not provide enough evidence for us to determine what is in Elizabeth's mind at this moment. Perhaps her gnomic response should be interpreted as an effort of self-distancing of the kind that has been attributed to Austen herself.

In a brief essay that was to inspire his longer work, *Jane Austen, or The Secret of Style,* D. A. Miller argues that the impersonal epigrammatic mastery evinced by Austen's style simultaneously strikes fear and love in her readers. It also eliminates the embodied specificity of the writer, for "her will to style is linked to the felt unrepresentability of her situation."[38] Why unrepresentable? According to Miller, social recognition of an individual in Austen's society required that individual to become legible within heterosexual marriage. Marriage, like style, is "a norm . . . at once inevitable and impossible" for someone who does not accommodate themselves to it.[39] "Jane Austen's manifestly ruthless style," Miller contends, "gains the novelist more tenacious devotion than any mere abusive parent, or withholding lover, could ever hope to inspire."[40] When Miller develops this argument in *Jane Austen,* he contextualizes it within the gender norms of the time. Someone like Robert Ferrars from *Sense and Sensibility* who violates both gender and sexual norms but "gets away with it" because of style becomes Miller's example: "Much better than Austen's manly heroes, then, the effeminate Robert teaches us the immense power, and the inestimable value, of an authority so sure of itself, so always-already taken for granted, that it doesn't even need the naturalizing alibi of a virile mission."[41]

Throughout this work on Austen, Miller assumes that she is a writer who wants to embody mastery but who, failing to do so, must perform it. I disagree. Take Elizabeth's quip as an example. Yes, it is a performance of mastery, but not in a consolatory move—not even, perhaps, in a personally defensive one. The first part of the sentence should be central to our understanding: "We all love to instruct." Teaching is not just about the learner, but about the pleasure gained by the teacher. Here the context is: we love to tell people what we think they really think. What about the second half of Elizabeth's sentence—the "not worth knowing"? We know what it means for something not to be "worth it": the result does not reward the effort. To teach something not worth knowing would be to ask students to expend precious energy learning something not worth the effort. We often think that all learning is good—all gaps should be filled—but here Austen suggests that is not the case.

The conjunction of knowledge and ignorance is heavily thematized at the end of *Pride and Prejudice,* and Elizabeth's aphoristic philosophizing—sometimes merged with the author-narrator's—abounds. In a moment in which the two Bennet sisters are coming to terms with their feelings, there is so much they still do not know. When Elizabeth and Darcy return to Longbourn after their engagement, Elizabeth is immediately asked by Jane and "all the others," "'My dear Lizzy, where can you have been walking to?'" She replies only "that they had wandered about, till she was beyond her own knowledge" (352). During the second proposal scene, Elizabeth and Darcy discuss the alterability of previous opinions:

> "But think no more of the letter. The feelings of the person who wrote, and the person who received it, are now so widely different from what they were then, that every unpleasant circumstance attending it, ought to be forgotten. You must learn some of my philosophy. Think only of the past as its remembrance gives you pleasure."
>
> "I cannot give you credit for any philosophy of the kind. *Your* retrospections must be so totally void of reproach, that the contentment arising from them, is not of philosophy, but what is much better, of ignorance. But with *me,* it is not so. Painful recollections will intrude, which cannot, which ought not to be repelled." (348–49)

Even in this moment of maximum intimacy, Elizabeth instructs ("You must learn") and Darcy erases not only her (playful) claim to expertise but her experience: he tells Elizabeth that her "philosophy" is unfounded. He does so, of course, in the form of a compliment, but, even so, he insists that the source of her pleasurable retrospections be "ignorance." It is Darcy here who wants to be the center of growth, who insists that he has had—if not in print—his own consciousness scenes.

Austen has funneled the reader's desire toward this very scene with Elizabeth and Darcy, a scene in which Elizabeth tries to teach him, and he refuses to learn. We could interpret that refusal as a sign of lingering masculine pride. I see in it, however, Austen's ongoing interest in ignorance as a result of failed instructions, and with failed instruction as not necessarily bad. Indeed, throughout *Pride and Prejudice* there is an erotic charge running through dynamics in which one person desires to teach and imagines the pleasure in instructing their beloved, while that beloved takes a perverse pleasure in refusing to be taught. Even Mary, after Elizabeth refuses Mr. Collins, is given the opportunity to entertain this fantasy of instruction in the only moment of free indirect thought she is permitted in the novel: "She rated [Mr. Collins's] abilities much higher than any of the others; there was a solidity in his reflections which often struck her, and though by no means so clever as herself she thought that if encouraged to read and improve himself by such an example as her's, he might become a very agreeable companion" (122). Perhaps what is so attractive about the desire-driven dynamic of instruction without learning is that, within its confines, ignorance becomes voluntary, and any party can step into the position of authority as often—or as infrequently—as they like.

Given the privileging of the sciences within the modern academy, there is pressure for literary scholars to demonstrate that literature can provide us with ways of knowing as well—with methods that run either parallel to or at odds with the methods of knowing employed in the sciences. What I am suggesting here is that Austen built into her fiction—fiction that was deeply indebted to what Hanlon calls "novels of data" and "novels of perception"—a thick description of ignorance as partial knowledge that departed from the ways in which these earlier novels thought about empirical learning. Furthermore, Austen does not dismiss this form of partial knowledge as useless because it is incomplete. Rather, both *Northanger Abbey* and *Pride and Prejudice* playfully imagine confident knowledge as useless and ignorance as useful in narrative terms.

Conclusion

AT THE END OF *Northanger Abbey,* Catherine Morland gets what she desires, and we readers get what we expect: marriage. Despite the General's initial fury at Henry, we perceive the dwindling pages of the novel we hold: "We are all hastening together to perfect felicity" (185). Concluding in nuptial bliss, Austen's novel epitomizes comfort for a particular kind of reader and leaves its critique of patriarchal values ambiguous.[1]

British Canadian novelist Rachel Cusk opens *Outline* (2014) in the wake of divorce, positioning it against the early nineteenth-century novel's drive toward coupling. Yet Austen's ownerless generalizations, aligned with both thought and (playful) ignorance, live on in Cusk's novel. *Outline,* the first work in the Outline Trilogy, is full of aphoristic observation. Every character dispenses anecdote and maxim. *Outline*'s first-person narrator is oddly neutral. Faye—unnamed until the last volume—perceives and responds to the world around her, but her true talent is listening to others talk. She records so much speech indirectly that it becomes difficult to discern where her reports of dialogue end and her internal monologue begins. Reviewers of Cusk's trilogy have focused on the role of the strangely passive Faye. While the relationship between gender and perspective is clearly elemental to this novelist's exploration of her form, it is overshadowed by the eerie universality of the many general propositions she delivers in novelistic prose.

From the first scene of *Outline,* we know this is a novel of storytellers and dispensers of wisdom. The Greek man sitting next to Faye on the darkened, airborne plane has a story to tell, one that comes in fits and starts and is ultimately delivered over a series of days. It is a story about multiple marriages, their endings and beginnings. Despite the necessary intrusion of wives and children, his story is also, strangely, about the birth and nurturance of a sentence. Eagerly he displays his respect for culture, and his belief that life's richness inheres not in the subtleties of experience, which are anyway embedded in particular times and places and thus easily forgotten, but in the subtleties of grammar and syntax.

Truisms pepper his discourse, as we can see from this bit of dialogue reported by the narrator early in her conversation with the man:

> A long time ago—so long that he had forgotten the author's name—he read some memorable lines in a story about a man who is trying to translate another story, by a much more famous author. In these lines—which, my neighbor said, he still remembers to this day—the translator says that a sentence is born into this world neither good nor bad, and that to establish its character is a question of the subtlest possible adjustments, a process of intuition to which exaggeration and force are fatal. Those lines concerned the art of writing, but looking around himself in early middle age my neighbor began to see that they applied just as much to the art of living.[2]

Meaning lies not in authorship but in "memorable lines" extracted from a translation multiply distanced from its origin. "A long time ago . . . he had forgotten . . . he read . . . he still remembers . . . a sentence." Cusk's novel takes as a principle of its form the question this book has been asking: What was—and what is—the role of general wisdom in realist fiction, even realist fiction of the contemporary variety, and does it have anything to do with the kind of situated, historical, and literary knowledge such novels ostensibly produce?

Cusk's Faye is a traveler, and yet "the only hope of finding anything," she tells us, "is to stay exactly where you are, at the agreed place. It's just a question of how long you can hold out" (13). The statement is in the present tense, not part of the novel's primarily retrospective narration. The "you" does not refer to a specific addressee, though the narrator has been delivering an anecdote about her son to that man sitting next to her on the plane. While this man need not do so, he is invited to stand in, as it were, for the second-person addressee. We readers of the novel are similarly invited to be addressed. We might expect that, as a realist novelist and memoirist producing literary fiction, Cusk would be interested in the testimonies of private experience. Her narrator's confessions, however, tend to conclude in oddly formal periods: "I said that I lived in London, having very recently moved from the house in the countryside where I had lived alone with my children for the past three years, and where for the seven years before that we had lived together with their father. It had been, in other words, our family home, and I had stayed to watch it become the grave of something I could no longer definitively call either a reality or an illusion" (11). No one really talks or reflects internally in the way Cusk's characters talk and reflect (it would be hard to stomach the pretension of a person who did), but many people write this way. Cusk is, in fact, interested in a mode of novelistic dialogue that is much more like writing than speaking. Her works are novelistic in that they are full of the details of private life, but these privacies are publicized not quietly, through free indirect discourse, but out loud within the fictional world. Her characters make very personal public declarations. The novel pretends that such testimonies can be understood by (almost) any reader, regardless of prior experience. Cusk disentangles story from storyteller so as to render story truly public in the sense of it being no longer privately owned. There are very few examples in our culture, Cusk suggests in an interview, in which "when you say something, you render it public and everyone owns it. It is no longer yours."[3] Yes, Cusk is playing with the idea that writers steal others' stories. Yet the form of each novel in the Outline Trilogy suggests that she does not adopt others' tales to render her fictions realer in a traditional sense; rather, she aims to render individual stories more universal, more the property of anyone.

Cusk depicts this formal experimentation with universality as a response to a realization she had about her own consciousness. She felt that her consciousness, what appeared to be "[her] individuality, [was] actually resting on old, possibly decrepit structures."[4] Ever the writer, Cusk decided to respond to this distressing realization at the level of the sentence in a way that plays with the boundaries between represented speech and thought. She describes the characters in her trilogy that deliver objective-sounding subjective accounts as "writing in inverted commas." They speak in writing. Alexandra Schwartz, interviewing Cusk for *The New Yorker*, asked if she intended Faye to be "the page on which" the many persons who speak to her are "writing." Cusk's response was curt and evasive: "Well, she's the only writer and she doesn't say anything." With that, Cusk transitioned into a meditation on speech that follows mention of the Miranda warning:

> That's the only sort of culturally available place in which, when you say something, you render it public and everyone owns it. It's rare that that's recognized. In our personal lives, when we tell someone something, we're really annoyed if they tell someone else. And, as a writer, that's a constant

pitfall because people talk and that's the life that's in front of you and it may well end up in your work. That is apparently a form of, not theft exactly, but of using real life. When I write a book, I don't feel I should decide who's allowed to read it. It's put out into space, and speaking is like that. That's partly what I'm trying to do in these monologues. I'm not interested in character because I don't think character exists anymore.[5]

Cusk references here the socially embedded characters of nineteenth-century realism. The things that "root[ed] to universality" in such works were people and places: "in the Victorian novel—the village, the vicar, the woman." When she sat down to write *Outline*, she felt those old things "were done with." The expression of life in a novel no longer needed to be determined so rigorously by identity and place. We now live more homogeneous lives. To seem real, experience needed to be narrated laterally and be less character determined. Following from this we might argue that, as individuals, we have little understanding of or property in our own experiences. There is much to discover about private life in others' words, but it is by nature an unsystematic mode of inquiry.

My contention is that we can understand Cusk's move toward the universal in language as a return to the unknowing maxim of the eighteenth-century novel rather than as a twenty-first-century innovation. This book has examined that unstudied tradition of maxims in early novels. Rather than applying a preexisting theory of the maxim to explain its function in eighteenth-century realist fiction, I have developed a new theory of the maxim embedded in a particular moment in time—the moment when this ancient rhetorical device was drawn into the currents, at times parallel, at times not, of empiricism and the early novel.[6] Nondidactic maxims flow through each chapter of this book, and, wherever they run, they sprout counterexamples to the "plain" representations of experience we are supposed to find in early science and the early novel.

Understandings of knowledge shifted in the early modern period. Knowledge was no longer to be general; it was now "contextual, specific, and historical."[7] With the advent of the Scientific Revolution, experience was to be controlled through experiment with the help of technology, and reason was to organize logically both the ideas derived from perception and inferences drawn from observation-based propositions. Whenever possible, internal processing of experience was to be rendered external, so that others could be invited to (modestly) witness it. *Maxims and the*

Mind has intervened in this narrative, tracing Francis Bacon's legacy not only to science but to the early novel. Bacon started with the premise that not-knowing was essential to science. His maximic "knowledge broken" was a format for transmitting knowledge in process. A collection of unordered maxims would encourage doubt and questioning more than unified treatises would. Yet Bacon's maxims provoked such doubt, in part, by satirizing the mind's natural restlessness and simultaneous desire for certainty and novelty. Broken knowledge was also a snapshot of the potential for the mind to remain lost in this darkened condition of unknowing. Such satirical generalizations were poignant and provoking, and even now, we literary historians treat and quote them as truths. La Rochefoucauld used a similar form of ironic, satirical maxim in an explicitly moral register. Both Bacon and La Rochefoucauld challenged the long-standing idea that self-evident, pithy statements foreclosed inquiry. Instead, their maxims direct readers toward horizons for future discovery. Novelists adopted such nontraditional maxims to communicate partially or imperfectly understood experiences belonging to themselves, their characters, or their narrators.

Jonathan Culler once wrote that theories of genre should leave room for a writer to get a genre wrong.[8] The writers I have featured were determined to get the maxim wrong. Whereas traditional maxims serve established truth and cultivate the feeling of shared understanding, self-canceling maxims provoke readers to question, doubt, and test received wisdom against experience. When the narrator of *Northanger Abbey*, for example, interrupts Catherine and Henry's courtship to give advice to her women readers about romantic attachment, she is stepping into the role of the self-consciously instructive writer, a role shaped by Austen's predecessors in the third-person novel (both Fieldings, Haywood, Burney) and in the moral essay (Steele and Addison; Haywood, again; and Johnson). Her predecessors dispense wisdom. Austen dispenses it as well, but with an irony that comes in the turn of her sentence. "To come with a well-informed mind," she archly informs readers, "is to come with an inability of administering to the vanity of others, which a sensible person would always wish to avoid."[9] Do not be foolish by insisting on being wise. Austen sees in the maxim, as Swift did, an opportunity for undermining entire dogmas or systems of knowledge. Writing almost a century after Swift, however, Austen has embraced the novel as a genre that can encourage antiauthoritarian thought. Where Swift used streams of sententiousness to warn readers of the personal perils of individual attempts at understanding experience, Austen encourages her readers to question the increasingly received wisdom of empirically inflected fiction.

Because of the ease with which maxims could be undermined within narrative, they became useful tools for authors wanting to portray an absence of thought or a confusion about the meaning of an event, especially when an event's public meaning conflicted with an individual's experience of it. When a character or narrator delivers a maxim whose general claim is contradicted by its particular context, that person ends up seeming naive or confused, and the writer has succeeded in expressing lack of understanding that would be otherwise difficult for a character herself to express. By gesturing at the limits and possibilities of fiction that aimed to seem true to many different readers' experiences across time, these maxims within realist fiction point toward alternative paths for the novel and for novel reading as a technique for understanding others' personal experiences and making sense of our own.

The maxims featured in this book work differently depending on whether they are inspiring a reader to question dominant philosophies (as in the case of Bacon's aphorisms, La Rochefoucauld's maxims and Behn's translation of them, or Swift's streams of sententiousness) or encouraging a reader to question the novel's methods of faithfully reproducing reality (as in the case of Samuel Richardson's literary generalizations or Austen's ironic maxims).

Bacon assumed that writer and reader would share the same goal regarding their objects of study. They should both want to understand how things work so they can manipulate cause and effect to achieve new results, deepening their understanding of physical mechanisms. A writer who chooses to communicate their ideas about science as a collection of aphorisms punctuated by gaps helps their readers suspend their judgment and turn to particulars before attempting to debate best practices. One goal of maxim-collection as a writing strategy is thus to prevent the act of scientific communication from being defined (and undermined) by ego: by the writer's desire to be accepted and the reader's desire to be smart.

The rhetorical situation is obviously different in the early novel. A writer of early realist fiction may want to teach, whereas a reader may want to be entertained. Another writer may want her personal account to seem true to her experience but ungeneralizable, whereas a reader may want to apply insights from that account to his own life. Additionally, a person who utters a familiar maxim need not take responsibility for the sentiment. In this way, maxims enable us to articulate personal experience indirectly.

Despite this difference between maxims in early science and maxims in the early novel, I believe we can learn something important about novels by taking seriously the possibility that writers and readers of early realist fiction sometimes shared the goal of understanding how living and thinking in the absence of full knowledge affects us. Self-canceling maxims in early novels illuminate this shared goal: the writers I examine frequently manipulate on the page traditional signs and figures of knowledge, from the maxim to the cultural objects that contain them (books) to the persons who most commonly deploy them (aristocratic men and later, middle-class women). The result is a series of new representations in nonreferential modes: a representation of the partially revealed secrets of love; of the mind struggling and failing to comprehend its body's actions; of the reality of publicly articulated experiences, the true meanings of which are only privately—and multiply—understood; and, finally, of the existence of commanding, feminine literary authority that nonetheless only offers knowledge not worth knowing.

If the goal of the New Science was ultimately human control of nature, the goal of the novel as presented here is not to control readers by constructing an ideologically powerful, totalizing version of self and world. Readers are also writers. Across time, the readerly writers I've featured in this book took seriously the partial, disconnected communications of their predecessors and turned to maxims in conjunction with textual particulars—in this case, the particulars of realist fiction, from its motifs of seduction and survival to its grammatical and syntactical markers of represented thought and perspective—to insist on the value of ignorance, partial knowledge, or incomplete understanding. There exists in the early novel a resistance to procedures we think of as empirical (particularized description; neutral narration). Yet this very mode of resistance, I show, has its origins in early science. Epistemological disrepair is not corrected, but rather harnessed, for the sake of the gradual process of knowledge production. And the more public and collective the process is, the better.

Two of the main writers and thinkers treated in the book—Aphra Behn and Jonathan Swift—work in the period before the novel's consolidation in the 1740s. Richardson, though he was part of that project, also questions the presentation of the feminine novelistic subject as a coherent idea against which other women should measure themselves.[10] Both Behn and Swift emphasize embodiment and the role of the passions over reason in human acts of perception, expression, and communication. Because experience is not dispassionate, neither believes in the possibility of transparent communication or of agreement over the meaning of collectively witnessed events. Life is rich, and knowledge is broken.

Later, for Richardson and Austen, novelistic conventions for representing thought become much more at issue. The representation of interiority is now part of what makes the experiences novels convey seem true to reality. Richardson's second novel, attuned to the disconnect between the experience of rape and its representation in the public sphere, reckons with the possibility that characters' and readers' experiences of what is true to reality might differ. How can a realist novel produce knowledge in a reader without access to that reader's own empirical and nonempirical "data sets"? A writer ultimately has no control over the sentences he produces, especially if he's borrowed them and they are general, leaving "no impression on the mind." Unimpressed, unknowing minds are free to say out loud what they read, and to mean what they say.

The maxim becomes a fascinating literary tool, given these interrelated issues. It is a form that, when plunged within a narrative context, can register incongruence between private experience of an event and the public articulation of what happened. By the end of the century, it also emerges as a form associated with feminine education and feminine ignorance. In Austen, Mary Bennet's publicly delivered maxims are not the negative image of her sister Elizabeth's consciousness scenes, but can be viewed instead as part of a parallel, alternative mode of feminine not-knowing, a mode in which articulations of one's partial knowledge do not have to open one up to the threat of potential social humiliation.

One idea to which this book repeatedly returns is that women writing in the long eighteenth century (including the fictional writer Clarissa Harlowe) found self-canceling maxims particularly appealing. Aphra Behn's early English translation of La Rochefoucauld's *Maximes*, for example, is, among other things, a deliberate display of a feminine philosophical authority newly associated with a materialist view of the universe. In merging this display of authority with embodied particularity, Behn marks her expressive act as both sexualized and scientific.

Knowledge production becomes a gradual and intimate act requiring both confidence and humility. Similarly, Richardson sees in the reader-writer dyad the potential for transmitting the truth of personal experience without having to put what is private into words. Richardson aligns the post-rape peculiarity of Clarissa's practice of literary generalization with her personal artifacts of scriptural devotion—artifacts that, once published by Clarissa (via her "editor," Richardson), could be easily adopted by readers for their own particular use. Although modern literary critics fear that the maxim, a form ostensibly representing self-evident truth or socially consented to fact, snuffs out or suppresses private experience that is credible only when it is formless, literary generalizations such as those in *Clarissa* move beyond this to suggest that public forms such as the maxim nonetheless remain charged with the possibility of bearing private experience. By the time Austen is writing, at the end of the eighteenth century and the beginning of the nineteenth, the maxim is more regularly interpreted as a feminine form, and the women who use it are marked as possessing not knowledge, but fragments of learning scavenged from others.

Although women writers found nontraditional maxims appealing, their uses of the form remained stubbornly associated with assumptions of feminine peculiarity, intellectual inferiority, and ignorance. Women writers attempting to navigate these associations from the seventeenth through the early nineteenth century could not win: if their applications of maxims were seen as too particular (as Behn's translation of La Rochefoucauld's maxims was taken to be), then they were judged as unable to grasp general principles. If they relied too much on general principles, as Mary Bennet does, then they were criticized for not attending to all the particulars of scholarship and observation of diverse human behavior over time.

We have tended to see interiority as the bedrock on which liberal subjectivity is built. To be a person—or "liberal subject"—one must possess that inner space of private subjectivity capable (at least in theory) of articulating personal truth untouched by social structures and norms. Structures and norms may infiltrate and shape this version of the self (as many argue is the case in Austen's works), but such common processes are nonetheless violations of the self's natural autonomy.[11] The novel form has been seen as both representing such socially compromised interiors and as helping readers celebrate a psychological complexity defined by a person's perceived difference from herself. Who would wish for such an uncomfortable alienation from world, self, and other? The instances of early fiction I've examined suggest the importance of not-knowing to early fictional figurations of inwardness. Existing at the border between the internal and external, maxims in these fictions focus attention away from the detailed accounting of thoughts and feelings toward an appreciation of the layers of incomprehension and the absence of understanding—in the novel and in science.

NOTES

Introduction

1. Daniel Defoe, *Robinson Crusoe*, ed. Thomas Keymer (Oxford University Press, 2007), 123.
2. Jonathan Swift, *Gulliver's Travels*, ed. David Womersley, vol. 16 of *The Cambridge Edition of the Works of Jonathan Swift* (Cambridge University Press, 2010), 416–17. Compare with the following maxim from Epicurus: "Thanks be to blessed Nature, that she has made what is necessary easy to obtain, and what is not easy unnecessary." Pierre Hadot, *Philosophy as a Way of Life: Spiritual Exercises from Socrates to Foucault*, translated by Michael Chase and edited by Arnold I. Davidson (Blackwell, 1995), 87.
3. Miguel de Cervantes Saavedra, *The Ingenious Hidalgo Don Quixote de la Mancha*, trans. John Rutherford (Penguin, 2000), 14. In the prologue, the novel's author bemoans to a friend that his book lacks "all erudition and instruction, [is] without any marginalia or endnotes, unlike other books ... that, even though they are fictional and not about religious subjects, are so crammed with maxims from Aristotle, Plato, and the whole herd of philosophers that they amaze their readers." This prologue does not appear in the second edition of Captain John Stevens's revised version of Thomas Shelton's *The History of the Most Ingenious Knight Don Quixote De la Mancha* (R. Chiswell, S. & J. Sprint, R. Battersby, S. Smith, B. Walford, M. Wotton, and G. Conyers, 1706).
4. Jane Austen, *Pride and Prejudice* [1813], ed. Vivien Jones (Penguin, 2014), 25.
5. John Bender, *Ends of Enlightenment* (Stanford University Press, 2012), 28.
6. See also Tita Chico's explication of Shapin and Schaffer's account of the "literary technology" necessary to the "virtual witnessing" that lent credibility to the matters of fact created through the process of experimentation and observation (*The Experimental Imagination: Literary Knowledge and Science in the British Enlightenment* [Stanford University Press, 2018], 18–19). Chico writes: "The association of 'realist representation' with early scientific writing (and Dutch painting) is suggestive, but presumes a stability and familiarity to what that might mean—not to mention that realism as a literary aesthetic emerges in the nineteenth century, not the seventeenth" (19).

7. Steven Shapin and Simon Schaffer, *Leviathan and the Air-Pump: Hobbes, Boyle, and the Experimental Life* (Princeton University Press, [1985] 2011), 60–65. Bender puts it this way: "Surrogate witnessing" is "the practice in early modern science of placing a single experiment at the foundation of a generalizing inductive process even though this unique experiment could not have been witnessed by the wide audience required for assent to newly defined general principles, or indeed witnessed by anyone or any but a very small group present at the experimental site" (*Ends of Enlightenment*, 28–29).
8. Bender, 4.
9. Helen Thompson, *Fictional Matter: Empiricism, Corpuscles, and the Novel* (University of Pennsylvania Press, 2016), 12–13; Kristin M. Girten, *Sensitive Witnesses: Feminist Materialism in the British Enlightenment* (Stanford University Press, 2024).
10. Robert N. Proctor, "Agnotology: A Missing Term to Describe the Cultural Production of Ignorance (and Its Study)," in *Agnotology: The Making and Unmaking of Ignorance,* ed. Robert N. Proctor and Londa Schiebinger (Stanford University Press, 2008), 2.
11. Brian Vickers, "Swift and the Baconian Idol," in *The World of Jonathan Swift,* ed. Brian Vickers (Basil Blackwell, 1968), 88–89.
12. See Michael McKeon, *The Origins of the English Novel, 1600–1740* (John Hopkins University Press, [1987] 2002), 66–68; and Thomas Keymer, introduction to *Robinson Crusoe,* by Daniel Defoe (Oxford University Press, [1719] 2007), xx, xxv.
13. Brad Pasanek, *Metaphors of Mind: An Eighteenth-Century Dictionary* (Johns Hopkins University Press, 2015), 208. Providing an example from Augustine, Pasanek writes that "inner and outer are used to structure distinctions between spirit and matter as early as the fourth century" (208).
14. Francis Bacon, *The New Organon,* ed. Lisa Jardine and Michael Silverthorne, trans. Michael Silverthorne (Cambridge University Press, 2000), 10. This quotation appears in the preface to *The Great Renewal* (also known as the "Great Instauration"), Bacon's deeply ambitious plan for a multipart, all-encompassing renewal of learning. This preface was included in the first edition of *The New Organon,* so Jardine and Silverthorne reproduce it in their translation. Elsewhere I will simply cite *The New Organon* when quoting from this preface.
15. Bacon, *New Organon,* 44–45.
16. Aphra Behn, "*The Fair Jilt; Or, the Amours of Prince Tarquin and Miranda,*" in *All the Histories and Novels Written by the Late Ingenious Mrs. Behn . . .* (R. Wellington, 1705), 142.

17. Jonathan Swift, "*Tatler* 5," in *The Writings of Jonathan Swift*, ed. Robert Greenberg and William Piper (W. W. Norton, 1973), 457. Steele's *Tatler* ceased on January 2, 1710/11. Less than two weeks later, there appeared the first number of another *Tatler* (known as *The Spurious Tatler* in the critical literature), created by William Harrison and Swift. Number five of *The Spurious Tatler* has been attributed to Swift. All subsequent references to this work will be included parenthetically.
18. Swift, "*Tatler* 5," 457.
19. John Locke, *An Essay Concerning Human Understanding*, ed. Peter H. Nidditch (Clarendon, 1975), 593.
20. Bacon, *New Organon*, 35, 37.
21. In "The Epistle to the Reader" that opens the *Essay*, Locke explains that initially he set down his "*hasty and undigested Thoughts*" as so many "*incoherent parcels.*" Only later did he, "*after long intervals of neglect*" was it "*brought into . . . order*" during a retirement for his health (7).
22. *Seneca Unmasqued* appears at the end of *Miscellany, Being a Collection of Poems by Several Hands* (J. Hindmarsh, 1685). See also Aphra Behn, *Seneca Unmasqued: A Bilingual Edition of Aphra Behn's Translation of La Rochefoucauld's* Maximes, ed. Irwin Primer (AMS, 2001). My subsequent references will be to Primer's modern scholarly edition. Primer demonstrates that Behn used La Rochefoucauld's fourth edition of the *Maximes* from 1675. For the history of the *Maximes*' English translations, see Joseph E. Tucker, "The Earliest English Translation of La Rochefoucauld's *Maxims*," *Modern Language Notes* 64, no. 6 (1949): 413–15.
23. The first version of this maxim I have quoted is from François de La Rochefoucauld, *Collected Maxims and Other Reflections*, trans. E. H. and A. M. Blackmore and Francine Giguère (Oxford University Press, 2007), V:111, 33. The Roman numerals followed by a colon and Arabic numerals denote the edition number and the numerical placement of the maxim in that edition. Behn's version of this maxim appears in Primer, *Seneca Unmasqued*, 369. Behn's numbering departs from the numbering in the fourth edition of the *Maximes*, with which she worked. In what follows I cite the page numbers and not the maxim numbers from *Seneca Unmasqued*. When I quote from the French, however, or when I provide other English transitions, I follow the Blackmore and Giguère edition, providing the Roman and Arabic numerals of the cited maxim for the reader's ease of reference and comparison to other editions of La Rochefoucauld's work.
24. Jonathan Swift, "Swift to Alexander Pope (Sep. 29, 1725)," in *The Writings of Jonathan Swift*, ed. Robert A. Greenberg and William Bowman Piper (W. W. Norton, 1973), 585.

25. Jonathan Swift, "Swift to Alexander Pope (Nov. 26, 1725)," in *The Writings of Jonathan Swift*, ed. Robert A. Greenberg and William Bowman Piper (W. W. Norton, 1973), 586.
26. Jonathan Swift, *The Battel of the Books*, in *The Cambridge Edition of the Works of Jonathan Swift*, vol. 1, *A Tale of a Tub and Other Works*, ed. Marcus Walsh (Cambridge University Press, 2010), 142.
27. Robin Valenza, "How Literature Becomes Knowledge: A Case Study," *ELH* 76, no. 1 (2009): 215–45.
28. Andrea Nightingale interprets Bacon's use of the phrase "broken knowledge" in this way, reading it as a condition to be overcome by science. Bacon's "broken knowledge" bears a close relationship to his "knowledge broken," which I discuss later in the introduction. For Nightingale, when Bacon evokes "broken knowledge" he is "denigrating wonder," an attitude he believed "scientists must repair . . . by the achievement of scientific knowledge." See Nightingale, "Broken Knowledge," in *The Re-Enchantment of the World: Secular Magic in a Rational Age*, ed. Joshua Landy and Michael Saler (Stanford University Press, 2009), 15–37, 15. Nightingale follows philosophers who advocate for apprehension of "nature (and our place in nature) in bodily and experiential terms" (16). This "ecological" approach accepts the partiality of the human perspective within a broader nonhuman ecology. It is both situated and unsituated.
29. This is also the trajectory that leads to modern aesthetic theory. See Abigail Zitin, *Practical Form: Abstraction, Technique, and Beauty in Eighteenth-Century Aesthetics* (Yale University Press, 2020), 28–30 and 37.
30. Samuel Richardson, *Clarissa, or, The History of a Young Lady*, ed. Angus Ross (Penguin, 1985), 985. This edition follows the text of the first edition published in 1747/48. All subsequent references to this edition of the novel will be abbreviated *C* and cited parenthetically.
31. Frances Ferguson, "Rape and the Rise of the Novel," *Representations* 20 (Autumn 1987): 88–112, 106.
32. Richardson, *Clarissa*, 893.
33. Swift, *Gulliver's Travels*, 416–17.
34. Marthe Robert, "From *Origins of the Novel*," in *Theory of the Novel: A Historical Approach*, ed. Michael McKeon (Johns Hopkins University Press, 2000), 57, 58.
35. Robert, "From *Origins of the Novel*," 58.
36. Ephraim Chambers, *Cyclopædia: or, an Universal Dictionary of Arts and Sciences* (James & John Knapton, John Darby, Daniel Midwinter, Arthur Bettesworth, John Senex, Robert Gosling, John Pemberton, William & John Innys, John Osborn & Thomas Longman, Charles Rivington, John Hooke, Ranew Robinson, Francis Clay, Aaron Ward, Edward Symon,

Daniel Browne, Andrew Johnson, and Thomas Osborn, 1728), 1:115, s.v. "aphorism." See also in the *Cyclopædia*, the definition of "MAXIMS, a kind of Propositions, which under the Name of *Maxims* and Axioms, have passed for Principles of Science; and which being self-evident, have been supposed innate" (512, s.v. "maxim").

37. Samuel Johnson, *A Dictionary of the English Language* (J. & P. Knaptor, T. & T. Longman, C. Hitch & L. Hawes, A. Millar, and R. & J. Dodsley, 1755), 1:148, s.v. "aphorism." Johnson defines "maxim" as "an axiom; a general principle; a leading truth" (2:120).

38. This truism emerged in twentieth-century accounts of the novel's eighteenth-century origins, and despite twenty-first-century revisions to the story of the early novel's celebration of the modern subject, these truisms remain active in work on the eighteenth-century novel.

39. Bender, *Ends of Enlightenment*, 41. Bender writes that "fictions, be they hypotheses or novels, yield a provisional reality, an 'as if,' that possesses an explanatory power lacking in ordinary experience."

40. See Sarah Tindal Kareem, *Eighteenth-Century Fiction and the Reinvention of Wonder* (Oxford University Press, 2014), 54–56. Kareem identifies three types of such techniques: linguistic, causal, and perspectival.

41. Roger Maioli, *Empiricism and the Early Theory of the Novel: Fielding to Austen* (Palgrave Macmillan, 2016).

42. Aaron R. Hanlon, *Empirical Knowledge in the Eighteenth-Century Novel: Beyond Realism* (Cambridge University Press, 2022), 2.

43. Peter Wehling, "Why Science Does Not Know: A Brief History of (the Notion of) Scientific Ignorance in the Twentieth and Early Twenty-First Centuries," in *Journal for the History of Knowledge* 2, no. 1 (2021): 1–13, 1.

44. Wehling, "Why Science Does Not Know," 1.

45. Bacon, *New Organon*, 38, 36.

46. Bacon, 18.

47. Robert Boyle, "Proëmial Essay," in *The Works of the Honourable Robert Boyle*, 2nd ed., ed. Thomas Birch, vol. 1 (London: J. & F. Rivington, 1772); quoted in Shapin and Schaffer, *Leviathan and the Air-Pump*, 66.

48. G. A. J. Rogers, "The Intellectual Setting and Aims of the *Essay*," in *The Cambridge Companion to Locke's "Essay Concerning Human Understanding*," ed. Lex Newman (Cambridge University Press, 2007), 7–32, 19–20.

49. Bacon, *New Organon*, 41, 42, 40.

50. Bacon, 42.

51. Bacon, 11. The Latin "indicia" has been translated to "directions."

52. Francis Bacon, *The Advancement of Learning*, in *The Major Works*, ed. Brian Vickers (Oxford University Press, 2008), 140. All further references to the *Advancement* of this edition will be abbreviated *A* and cited parenthetically.

53. R. W. Serjeantson, "Proof and Persuasion," in *Early Modern Science*, ed. Katherine Park and Lorraine Daston, vol. 3 of *The Cambridge History of Science* (Cambridge University Press, 2006), 140.
54. Rhodri Lewis points out that these "Arts Intellectual" roughly resemble the "parts" of oratory, with invention and tradition corresponding to *inventio* and *dispositio*. Such echoes between the rhetorical and logical arts are, however, not at all unusual in the period and should not lead us to believe that either invention or the bulk of tradition had a primarily rhetorical cast for Bacon. Indeed, Bacon places rhetoric at the very bottom of this hierarchy of intellectual arts: he lists it as the third of tradition's three constituents. Rhodri Lewis, "A Kind of Sagacity: Francis Bacon, the *Ars memoriae*, and the Pursuit of Natural Knowledge," *Intellectual History Review* 19, no. 2 (2009): 156. On the cross-pollination of logic and rhetoric throughout the early modern period, see Quentin Skinner, *Reason and Rhetoric in the Philosophy of Hobbes* (Cambridge University Press, 1997).
55. Lisa Jardine argues that Bacon was indebted to Renaissance debates about dialectic for both his opposition to Aristotelian logic and his adaptation of it. Jardine, *Francis Bacon: Discovery and the Art of Discourse* (Cambridge University Press, 1974). Humanist dialecticians such as Agricola, Melanchthon, and Ramus attempted to systematize and simplify the curriculum for the benefit of nonspecialists. This meant doing away with the classical distinction between certain or *demonstrable* series of inferences and merely *dialectical* or plausible reasoning suitable for debate. Such "debates," however, are restricted for Jardine to those occurring within and around dialectic handbooks. She argues that "tempting as it is to see Bacon's works (as some critics have done) as a development of the sophisticated reassessment of Aristotle's Organon by the schools of Oxford, Paris, and Padua, and hence to invoke a pat scholastic background, it is in fact possible to account both for Bacon's 'Aristotelian' assumptions and for his 'anti-Aristotelian' polemic in terms of the content of the dialectic manual, and contemporary polemical discussions of dialectical method" (10).
56. In the *Advancement*, Bacon records the attempts of these reformers and complains that they overcorrected for the convoluted Scholastic schemes of inference. Radical Renaissance dialecticians—Ramus chief among them—confused methods meant to *discover* new principles with methods designed merely to *teach* them. This point is at the center of Jardine's study: Bacon's "obsession," as she puts it, with "procedure of discovery and methods of presentation" (7).
57. For Bacon's definition of scientific as opposed to rhetorical "invention," see Bacon, *Advancement*, 222–23: "The invention of speech or argument

is not properly an invention: for to invent is to discover that we know not, and not to recover or resummon that which we already know."
58. Bacon negatively employs the term "magistral" in book 1. In that earlier passage, methods "magistral and peremptory" are contrasted with those "ingenuous and faithful," the former delivering knowledge "in a sort as may be soonest believed, and not easiliest examined," appropriate only in "compendious treatises for practice" (*A* 147).
59. "Probation," n. OED Online (accessed January 15, 2022).
60. Vickers, ed., *Major Works*, 642n233.
61. Francis Bacon, *Of the Advancement and Proficiencie of Learning: or the Partitions of Sciences*, trans. Gilbert Wats (Thomas Williams, 1674), 176.
62. Rhodri Lewis, "Francis Bacon, Allegory and the Uses of Myth," *Review of English Studies* 61, no. 250 (2010): 360–89. Lewis writes: "For Bacon, in other words, any text that purports to impart true learning must do so heuristically, thereby initiating the student into the true significance of what was being taught" (369).
63. "Systematic exposition" is Vickers's gloss. Lisa Jardine explains this passage as a direct allusion to Ramus's "dichotomizing method." See Jardine, *Francis Bacon*, 175n3.
64. Lisa Jardine identifies aphorizing's mixed quality of incompleteness united with control as the key feature of Baconian aphorisms, which are "designed to indicate that this is 'work in progress', susceptible of improvement and refinement," while also revealing "the competence of the author." Jardine, introduction to *The New Organon*, by Francis Bacon, ed. Lisa Jardine and Michael Silverthorne (Cambridge University Press, 2000), xviii.
65. Rhodri Lewis argues that Bacon saw receivers of information conveyed probatively as "heuristic and creative [agents], completing [the work's] field of reference for himself through an act of rational reconstruction," "Francis Bacon," 381.
66. Ian Watt, *The Rise of the Novel: Studies in Defoe, Richardson, and Fielding*, 2nd ed. (University of California Press, [1957] 2001), 29.
67. Watt, *Rise of the Novel*, 28.
68. Aristotle's *Organon* comprises his six works on logic: the *Categories*, *On Interpretation*, *Prior Analytics*, *Posterior Analytics*, the *Topics*, and *Sophistical Refutations*. Bacon's starting point for his *Novum organum* was the induction mentioned by Aristotle in the *Posterior Analytics*. Bacon argued that with his *Novum organum* it was possible to "perform such an induction" in a new manner; "that is, to infer the general rule from the particulars in which it inheres." Jardine, *Francis Bacon*, 6.
69. Book 1 of the *Novum organum* deals extensively with the famous "Idols" of the mind and the previous causes of errors in the sciences. Lisa Jardine

and Rhodri Lewis have both commented that Bacon's use of aphoristic delivery only approaches the probative method in his scientific works; it is in the service of a magistral or dogmatic method that aphorisms are employed in his *Essays*. See Lewis, "Allegory and the Uses of Myth," 368–69; Jardine, *Francis Bacon*, 174–78. Jardine goes furthest here in declaring that the *Novum organum* is not even itself probative. She argues that "Bacon's stringent requirements for the initiative [i.e. probative] method mean that all his own writings (with the possible exception of the fragmentary histories of life and death, winds, dense and rare, etc.) are themselves excluded from this category" (175).

70. *The Philosophical Works of Francis Bacon . . . Methodized, and made English, from the Originals, with Occasional Notes, To Explain what is Obscure*, 3 vols., ed. and trans. Peter Shaw (J. J. & P. Knapton, D. Midwinter & A. Ward, A. Bettesworth & C. Hitch, J. Pemberton, J. Osborn & T. Longman, C. Rivington, F. Clay, J. Batley, R. Hett, and T. Hatchett, 1733), 2:345.
71. Bacon, *New Organon*, 29.
72. Bacon, 33.
73. Bacon, 27.
74. Bacon, 28, 30.
75. Sandra Macpherson, *Harm's Way: Tragic Responsibility and the Novel Form* (Johns Hopkins University Press, 2010), 5.
76. Macpherson, *Harm's Way*, 6. "Subjective possibility" is from Franco Moretti, *The Way of the World: The Bildungsroman in European Culture*, trans. Albert Sbragia (Verso, [1987] 2000), 45–46.
77. Bacon, *New Organon*, 12.
78. Bacon, 18.
79. Bacon, 19.
80. Michael McKeon presents us with a related version of a conflicted Bacon when he writes: "The Baconian scientific program contains two contrary movements. An optimistic faith in the power of empirical method to discover natural essences points in one direction; a wary skepticism of the evidence of the senses and its meditating capacity points in quite another" (*Origins of the English Novel*, 68).
81. Kristin M. Girten, "Mingling with Matter: Tactile Microscopy and the Philosophic Mind in Brobdingnag and Beyond," *Eighteenth Century* 54, no. 4 (Winter 2013): 497–520, 500. In her footnotes, Girten cites Abraham Cowley's "Ode to the Royal Society," in Thomas Sprat's *History of the Royal Society* [1667], ed. Jackson I. Cope and Harold Whitmore (Washington University Studies, 1958), 66–73.
82. Girten, "Mingling with Matter," 500. Girten goes on: "Regarding Bacon's idols, Daston and Galison remark, 'only one of the four categories (the

idols of the cave) applied to the individual psyche and could therefore be a candidate for subjectivity in the modern sense (the others refer to errors inherent in the human species, language, and theories, respectively).'" Girten helpfully acknowledges that Bacon's "notion of the self differs in significant ways from more recent notions of subjectivity," she nonetheless argues that he "portrays all of the idols as having a negative impact on individual men's psychological experience and perception" (500–501). With the phrase "view from nowhere" Girten is referencing Thomas Nagel's *The View from Nowhere* (Oxford University Press, 1986).

83. Girten explores what she calls a "tactile empiricism," which "entails a distinctive interpenetration of subject and object" (501). In her argument, Swift's satire on the new philosophy calls attention to this "interplay between touch and sight," suggesting "the propensity of empiricism to facilitate intimacies between philosophers and their objects of investigation via this interplay—intimacies that, as a result of the interpenetration (not to mention excitement) they entail, contradict the reserve expected of adherents to a Baconian scientific method" (508). In other words, tactile empiricism—the perceptual interrelatedness of touch and vision—conflicts with Baconian empiricism's demand for a separation of the human mind from material nature. Bacon, according to Girten, "envision[s] the creation of a boundary between the self and its objects of study. . . . When he envisions a mind 'freed and cleansed,' he envisions a psyche that has risen above both its own preconceptions or biases as well as the material world." Girten argues: "The treatment of perception that John Locke offers in his *Essay Concerning Human Understanding* (1689) corroborates Hooke's portrayal of the microscope as both a visual and tactile instrument. The interrelatedness of touch and vision is, of course, the subject of the famous Molyneux Question to which Locke responds in the second edition of his *Essay* (1694)" (502).

84. Pasanek, *Metaphors of Mind*, 210.
85. Pasanek, 216.
86. Swift, *Battel of the Books*, 152.
87. Marcus Walsh, in his Cambridge edition of the *Battel*, gives the example of an earlier appropriation of Bacon's metaphor to a similar use. See Walsh 475–76n50. See also Robert Boyle's use of the spider as an image for those who fail to apply their faculties to a careful investigation of the universe (475–76n50). For more on the sources of Swift's spider and bee episode, see R. F. Borkat, "The Spider and the Bee: Jonathan Swift's Reversal of Tradition in *The Battle of the Books*," ECL, no. 3 (1976): 44–46.
88. Swift, *Battel of the Books*, 152.
89. Macpherson, *Harm's Way*, 175.

90. Macpherson, 24.
91. Stephanie Insley Hershinow, *Born Yesterday: Inexperience and the Early Realist Novel* (Johns Hopkins University Press, 2019), 8, 6. Of all of these studies, Hershinow's is perhaps the one that addresses most directly the idea of the *knowledge* novels produce in an empirical register and the *ignorance* they prize.

1. "Odd Fantastick Maxims"

1. The first English translation was *Epictetus Junior, or Maximes of Modern Morality* (T. Bassett, 1670), by J.D. of Kidwelly. In Kidwelly's translation, La Rochefoucauld's Epicureanism or Augustinian anti-Stoicism is muted and the work turned into more of a conduct manual. For example, what Behn translates as "As 'tis the character of a great Wit to express much in a few words, so 'tis of a little wit, to talk much to little purpose" (*Seneca Unmasqued*, 34), Kidwelly gives as "True Eloquence consists in saying whatever is requisite, and in not saying any more then what is requisite" (*Epictetus Junior*, 29). Kidwelly praises the author of these "Maximes of Modern Morality," who departs from the "*Casuisme*" that has "of late years" so "pester'd" morality (vi). Under these "impure hands" morality has been "disorder'd and debauch'd into pestilent and pernicious deductions (vii)." By contrast, these modern maxims display both "*Excellency* and *Novelty*" (vii). Kidwelly's dedication makes clear the anti-Scholasticism of what he calls "modern morality." Ironically, by the mid-eighteenth century, La Rochefoucauld's *Maximes* had generated the opinion among pious English writers that no other work had inspired in its readers such a hatred for humankind. This is, in fact, what the writer of the preface to the 1755 edition of Richardson's *Collection of Moral and Instructive Sentiments* contends. According to the prefacer, the translator of the *Maximes* supposedly says, "I have not read any thing in this age, that has given me a greater *contempt* for man." Samuel Richardson, *A Collection of the Moral and Instructive Sentiments, Maxims, Cautions, and Reflexions, Contained in the Histories of Pamela, Clarissa, and Sir Charles Grandison* (Samuel Richardson, 1755), viii.
2. The French reads "On a bien de la peine à rompre, quand on ne s'aime plus." La Rochefoucauld, *Collected Maxims*, V:351, 96.
3. Personal secrets are not the same as private thoughts. Michael McKeon provides a helpful distinction between secrecy and privacy: "Secrecy involves an act of 'intentional concealment,' whereas privacy, a central principle of negative liberty, names 'the condition of being protected from unwanted access by others.' . . . Secrecy is first of all a category of

traditional knowledge, not a privative privacy but that which distinguishes an elite from the deprived majority and which paradoxically comes to be fully known as secret only under the apprehension of its discovery." McKeon, *The Secret History of Domesticity: Public, Private, and the Division of Knowledge* (Johns Hopkins University Press, 2005), 469; McKeon quotes Sissela Bok, *Secrecy: On the Ethics of Concealment and Revelation* (Pantheon, 1982), 5–14.

4. Experimental philosophers had to be appropriately trained in techniques of (neutral) observation. Through training their minds would become suitably "modest," a moral condition of mental purity that would extend to the written records of the philosophers' observations. This account of the modest witness originates with Shapin and Schaffer's *Leviathan and the Air-Pump*. Drawing on Shapin and Schaffer's work, Tita Chico contends that a modest observer adheres to a moral value of abandoning the distractions of interest: "He embodies, paradoxically, a privileged absence, a model of spectatorship that conveys authority through its claim to be free from the limitations of self-interest" (*Experimental Imagination*, 37).

5. Girten, *Sensitive Witnesses*, 76.

6. Aphra Behn, *Seneca Unmasqued: A Bilingual Edition of Aphra Behn's Translation of La Rochefoucauld's* Maximes, ed. Irwin Primer (AMS, 2001). Primer's bilingual edition is particularly useful in its careful noting of the numbers from the fourth edition of La Rochefoucauld's *Maximes* to which Behn's translated maxims correspond. All other references to this edition will be abbreviated *SU* and cited parenthetically. *Seneca Unmasqued* was first published in *Miscellany, Being a Collection of Poems by Several Hands. Together with Reflections on Morality, or, Seneca Unmasqued* (J. Hindmarsh, 1685). Most modern editions of La Rochefoucauld's maxims reproduce the text from the fifth edition published in 1678. For her translation Behn used the fourth edition of 1675. A later translator of La Rochefoucauld's *Maximes* and competitor of Behn criticized her work on the basis of her departure from the original ("Mrs. Behn . . . seems not to have intended a perfect Work, so much as Entertaining her Self and her *Lysander*, with such Passages as were most applicable to her darling Passion of *Love*. Upon which occasion and some others, she takes the Freedom of Paraphrasing, and Accommodating as she saw fit, more perhaps to her own Diversion, than the doing Justice to the Author" [quoted in Primer, *Seneca Unmasqued*, xxi]). Such practices of imitation as opposed to strict translation were common, however, and Behn additionally "borrows" elements of La Chapelle-Bessé's preface to an early edition of the *Maximes*. See Primer, introduction and appendix 2.

7. See Primer, "Introduction," *Seneca Unmasqued,* xv–xvi; and Janet Todd, *The Secret Life of Aphra Behn* (Rutgers University Press, 1997), 374.
8. Janet Todd provides some context for this dedication, writing: "Much of the jolksy, seemingly intimate address to Lysander in her dedication was actually an unacknowledged translation of a French foreword to a 1665 [sic] edition of La Rochefoucauld. This tried to place the *Maximes* in a Christian context in which self-love was also condemned, and was probably added to counter church criticism of its pervasive cynicism. Behn omitted the Christian placing, but kept the familiar tone." Todd, *Secret Life,* 373.
9. Behn, *Seneca Unmasqued,* xlvii. As Irwin Primer points out in his edition of *Seneca Unmasqued,* Behn is directly translating and borrowing here from La Chapelle-Bessé's *Discours sur les Réflexions ou Sentences et Maximes morales,* the preface originally appearing in the authorized October 1664 edition of La Rochefoucauld's *Maximes* (Primer calls this the 1665 edition). Yet while Behn borrows from La Chapelle-Bessé here and elsewhere, her language is distinct and unique. Nowhere in La Chapelle-Bessé's scholarly preface, for example, do we find the French equivalent of "Idleness" or of the phrase "Lazy Diversion." For an exhaustive comparison that illuminates Behn's borrowings and additions, see Primer, *Seneca Unmasqued,* 140–41.
10. Primer, 162n31.
11. It is Behn and not La Chapelle-Bessé, from whose own French preface to the fourth edition she draws, who compares to lovers persons writing to ease their minds.
12. On Behn's use of the Lockean term "uneasiness" to describe a feeling inextricable from a mental cause of action, see Jonathan Kramnick, *Actions and Objects from Hobbes to Richardson* (Stanford University Press, 2010), 163.
13. This point is made by Pierre Force in a study that draws on thinkers such as La Rochefoucauld, Bayle, Mandeville, Hume, Montesquieu, Rousseau, and Smith in order "to explain how one goes from the interest doctrine (selfish motives are behind all human actions) to economic science (self-interest explains *economic* behavior, but not all types of human behavior)." Force, *Self-Interest Before Adam Smith: A Genealogy of Economic Science* (Cambridge University Press, 2003), 4.
14. This is cited by E. H. and A. M. Blackmore and Francine Giguère in their introduction to *Collected Maxims and Other Reflections,* François de La Rochefoucauld, ed. and trans. E. H. and A. M. Blackmore and Francine Giguère (Oxford University Press, 2007), xiii–xiv.
15. For background on "the Augustinian anti-Stoic polemic in seventeenth-century France," including "the more incisive psychological criticisms developed by Blaise Pascal, Nicolas Malebranche, and La Rochefoucauld"

(xvi), see Christoper Brooke, *Philosophic Pride: Stoicism and Political Thought from Lipsius to Rousseau* (Princeton University Press, 2012), 76–100. There are also those who identify an Epicurean strain in La Rochefoucauld's work. According to such arguments, his account of "virtue" as hypocritical emphasizes pleasure and pain as the driving principles behind human behavior.

16. Most of what we know about Stoicism, which originated in Athens around 300 BC, comes from three major figures whose works reemerged during the European Renaissance: Seneca, Epictetus, and Marcus Aurelius.
17. Hadot, *Philosophy as a Way of Life*, 83.
18. Hadot, 86.
19. With the help of Paul, Augustine argues that the division of body and mind does not map perfectly onto the division of flesh and spirit. Thus, we can include certain "vices of the mind" within the category of what Paul calls "works of the flesh" (Brooke, *Philosophic Pride*, 4). We might consider Augustine's conception of the materiality or fleshliness of the fallen intellect alongside Bacon's concerns about human intellectual impurity, concerns that led him to emphasize the need for mental modesty.
20. La Rochefoucauld, *Collected Maxims*, V:1, 3. The Roman numerals followed by a colon and Arabic numerals denote the edition number and maxim number provided by La Rochefoucauld. Behn's numbering departs from the numbering of maxims in the fourth edition, with which she worked. On Augustinian anti-Stoicism, see Brooke, xiv.
21. *All the Works of Epictetus, Which Are Now Extant; Consisting of His Discourses . . . In Four Books*, trans. Elizabeth Carter (S. Richardson, 1758), ii.
22. La Rochefoucauld, *Collected Maxims*, V:6, 5.
23. La Rochefoucauld, V:8, 5.
24. La Rochefoucauld was "in almost daily contact with Madame de La Fayette during the years when she was writing *La Princesse de Clèves* (1672–77); thus he may well have influenced that novel—and, conversely, she may well have influenced the last three editions of his maxims" (Blackmore and Giguère, "Introduction," xv.) See also Marie-Madeleine Pioche de La Vergne, Comtesse de Lafayette, *Zayde: A Spanish Romance*, ed. and trans. Nicholas D. Paige (University of Chicago Press, 2006).
25. Terence Cave, "Introduction," in *La Princesse de Clèves*, trans. Terence Cave (Oxford University Press, 1992), xi.
26. Nicholas Paige, *Before Fiction: The Ancien Régime of the Novel* (University of Pennsylvania Press, 2011), 39.
27. Paige, *Before Fiction*, 44.
28. Paige, 39. A concrete example from the novel helps Paige make this point. One popular seventeenth-century record of noble exploits during the French Renaissance reveals not only that the de Clèves brother whom

Mademoiselle de Chartres marries in the novel was never married in reality; this record also exposes the fact that the real Nemours—the Princess's secret love interest in the novel—really did compete wearing the colors yellow and black in the tournament described in detail in Lafayette's novel. Yet, as Paige describes, Valincour is horrified to discover that in reality "Nemours chose the colors of a woman whose sexual favors he was then enjoying" (41). In the novel, of course, the Princess is rigorously virtuous, perhaps to a fault.

29. Paige, 37.
30. Behn places "Of LOVE" alongside two other special sections, "SELF-LOVE" and "Of DEATH," all of which appear at the end of her collection. "Of LOVE" runs from *Seneca Unmasqued*, 104–28 (maxim nos. 313–90), "SELF-LOVE" appears at *Seneca Unmasqued*, 128 (nos. 390–93), and "Of DEATH" from *Seneca Unmasqued*, 128–34 (nos. 394 and 395) in Primer's edition of *Seneca Unmasqued*.
31. La Rochefoucauld, *Collected Maxims*, V:395, 109, emphasis mine.
32. La Rochefoucauld, V:111, 33, emphasis mine.
33. La Rochefoucauld, V:374, 103.
34. La Rochefoucauld, V:123, 36–37.
35. Primer, "Introduction," xxix.
36. Primer, xxviii.
37. Roland Barthes, "La Rochefoucauld: 'Reflections or Sentences and Maxims,'" in *New Critical Essays*, trans. Richard Howard (Hill & Wang, 1980), 3.
38. "Interest," n. 5, OED Online (accessed November 11, 2021). Compare to "self-interest," n.2.: "Preoccupation with, or pursuit of, one's own advantage or welfare, esp. to the exclusion of consideration for others" (accessed November 7, 2021).
39. The French reads: "L'intérêt parle toutes sortes de langues, et joue toutes sortes de personnages, même celui de désintéressé" (La Rochefoucauld, *Collected Maxims*, V:39, 14).
40. "The Publisher to the Reader," in La Rochefoucauld, 3. This was published as a preface to the fifth edition of the *Maximes* in 1678.
41. The French reads: "Il n'y a point de passion où l'amour de soi-même règne si puissamment que dans l'amour; et on est toujours plus disposé à sacrifier le repos de ce qu'on aime, qu'à perdre le sien" (V:262, 74).
42. Todd, *Secret Life*, 370–72.
43. Ross Chambers, *Story and Situation: Narrative Seduction and the Power of Fiction* (Manchester University Press, 1984), 13; quoted in Ros Ballaster, *Seductive Forms: Women's Amatory Fiction from 1684 to 1740* (Clarendon, 1992), 24–25.
44. Bacon, *Advancement of Learning*, 233.

45. Behn was among others who contributed "commendatory verse" prefacing Thomas Creech's 1682 translation of the Roman Epicurean poet Lucretius's *De Rerum Natura*. See Alvin Snider, "Atoms and Seeds: Aphra Behn's Lucretius," *CLIO* 33, no. 1 (2003): 1–24. On women writers and the Epicurean revival, see also Richard Kroll, *The Material World: Literate Culture in the Restoration and Early Eighteenth Century* (Johns Hopkins UP, 1991), 27, 85–111.
46. See Snider, "Atoms and Seeds." Snider writes that within Lucretian Epicureanism "sexuality occupies an intermediary position between the physical and the mental realms, a middle ground between bodily function and erotic reveries. Behn well understood that Lucretius ... regarded desire as an uncontrollable physical urge, much like hunger or thirst. Inverting the gendered perspective of the Latin poet and his translator, but also refusing to idealize libidinal drives as tender affection, Behn draws on both ancient and early modern tropes to make poetic invention the servant of eros, mechanics the basis of metaphysics, and the body a bundle of reflexes and emotions" (5–6).
47. This account of the modest witness originates with Shapin and Schaffer's *Leviathan and the Air-Pump*. Donna Haraway later expands upon it in her own critique. See Donna J. Haraway, *Modest_Witness@Second_Millennium.FemaleMan©Meets_OncoMouse™: Feminism and Technoscience* (Routledge, 1997), 23–30.
48. Chico makes the important point that "the observers were not necessarily the experimenters," especially as most Royal Society members "relied on men of lower rank" to perform the experiments that they observed and recorded (*Experimental Imagination*, 37).
49. Chico, 37.
50. Bacon, *New Organon*, 33.
51. One might be tempted to describe the difference between Bacon and Behn as the difference between an Augustinian approach to knowledge and labor and an Epicurean approach. To the Epicureans, self-interest is *the* first principle of human behavior because of the need to survive; for Rousseau in the eighteenth century, for example, self-interest is not a perverse consequence of modern commercialism, but a quality necessary for survival. The difference between the Epicurean and the Augustinian view here is that the latter saw self-love as a product of the Fall—a reorientation of the love that would otherwise naturally be directed toward God.
52. Joshua Foa Dienstag, *Pessimism: Philosophy, Ethic, Spirit* (Princeton University Press, 2006), 229–30.
53. Behn identifies in "the Duke's" maxims an "Air Gallant," pointing to a gallantry associated with men of wit and fashion, but also amorous men attentive to women. And yet Behn's characterization of the author of these

French maxims as noble and gallant aligns with a standard of leisure and privacy that will later become associated with femininity.
54. Sarah Raff, "Quixotes, Precepts, and Galateas: The Didactic Novel in Eighteenth-Century Britain," *Comparative Literature Studies* 43, no. 4 (2006): 466–81, 472.
55. Raff, "Quixotes, Precepts, and Galateas," 473.
56. Raff, 473.
57. Madame de Lafayette, *La Princesse de Clèves*, trans. Terence Cave (Oxford, 1992), 3.
58. Lafayette, *Princesse de Clèves*, 15.
59. For Paige, Lafayette's *Princesse* is not the harbinger of modern, third-person omniscient narrative fiction, which it is so frequently taken to be. Rather, it is "an isolated manipulation of longstanding conventions and local practices that changed precisely nothing" (*Before Fiction*, 36).
60. Paige, 38.
61. Paige, 48.
62. Paige, 50.
63. "Lafayette's characters . . . are not so much examples of amorous temperaments that one can laud or chastise as they are people defined by a fundamental inadequation between interior and exterior, feelings and actions. And Valincour signals that inadequation as the source of our identificatory pleasure" (Paige, 51).
64. La Rochefoucauld, *Collected Maxims*, V:374, 103.
65. According to Karen Newman, in her essay on Bonnecorse's *La Montre d'Amour*, "Behn's text follows Bonnecorse's closely, often word for word, but she introduces substantial new material as well." Newman, "'Wit's Great Columbus': Aphra Behn Translates *La Montre*, or *The Lover's Watch*," *Shakespeare Studies* 48 (2020): 101–7, 102. Newman's focus is not on Behn's translation itself but "the series of letters and poems by admiring contemporaries that preface Behn's text and which offer an interesting take on contemporary attitudes toward translation" (103). Newman emphasizes how the prefatory material for *The Lover's Watch* portrays Behn's translation as creation as opposed to imitation. I have not been able to locate the 1666 or 1671 French edition of Bonnecorse's work with which to compare Behn's translation, so I cannot tell whether Bonnecorse's original included the La Rochefoucauld maxim or whether Behn added it. In terms of publication chronology, either could be the case.
66. The two translations—*La Montre* and *Seneca Unmasqued*—are thus connected beyond their French originals. In the introduction to his edition of Behn's *Seneca Unmasqued*, Irwin Primer surveys late-seventeenth-century translation theory and outlines Behn's own approach to translation, which,

following Dryden, he identifies as neither word-for-word nor paraphrase but "imitation" ("Introduction," xx–xxiv). For more on Behn's approach to translation, particularly from French, see her "Essay on Translated Prose," the preface to her translation of Fontenelle's *Entretiens sur la Pluralité des Mondes* (1686), which she titles *A Discovery of New Worlds* (1688).
67. Aphra Behn, *The Lover's Watch*, in *The Works of Aphra Behn*, ed. Montague Summers (Benjamin Blom, [1915] 1967), 6:39.
68. Behn, *Lover's Watch*, 39.
69. Behn, *Seneca Unmasqued*, 110. The publication dates suggest to me that Behn may have added this maxim of La Rochefoucauld's, and that Bonnecorse did not include it in his original. It is, of course, possible that Bonnecorse was quoting from an earlier edition of the *Maximes*, which would have included this maxim on Love. But would Behn have recognized it and taken the trouble to go back to her translation from the previous year to copy it exactly (with those distinctive "'em"s)? It seems more likely that Behn would have been motivated to go back to her previous work if she wanted to make a particular point about Love and secrecy, one that was very much still in her mind after translating La Rochefoucauld.
70. Behn, *Lover's Watch*, 39.
71. Behn, 39–40.
72. See Mannheimer on how Behn negotiates the "public" spectacle of the theater and the "private" act of reading protonovelistic prose fiction. Katherine Mannheimer, "Celestial Bodies: Readerly Rapture as Theatrical Spectacle in Aphra Behn's *Emperor of the Moon*," *Restoration* 35, no. 1 (2011): 39–60.
73. Susan Staves, *A Literary History of Women's Writing in Britain, 1660–1789* (Cambridge University Press, 2006), 78. Staves sees as a contradiction the coexistence of the narrative's elevation of lovers willing to lose all and the narrator's moral condemnation of criminality.
74. Mannheimer, discussing Behn's navigation of shifting epistemological paradigms of performance and prose fiction in the 1680s, argues that *The Emperor of the Moon* "[casts] doubt on the culturally-assigned 'gender' of immersive reading" and "begins to undermine the distinctions between private and public, personal and political, mind and body" ("Celestial Bodies," 42). In Behn's *The Emperor of the Moon*, Baliardo attempts to use his telescope to plumb the secrets of the lunar king's closet, and thereby "evokes all too clearly the prototypical consumer of prose fiction: a figure already recognizable in the late seventeenth century, and already identified as female. Yet by embedding such imaginative transport within an expansive theatricality, Behn causes this 'feminized' form of reading to become public" (42).

2. The Maxims of Swift's Psychological Fiction

1. Swift, *Gulliver's Travels*, 437. All further references in this chapter are to this edition and will be abbreviated *GT* and cited parenthetically.
2. Virgil *Aeneid* 2.79–80. The cited translation is provided by Womersley in his editorial notes (*GT* lxvii n. 107, 437 n. 6).
3. Michael McKeon discusses Gulliver's naive empiricism in *Origins of the English Novel*, 348.
4. Jonathan Swift, *A Tale of a Tub*, in *A Tale of a Tub and Other Works*, ed. Marcus Walsh, vol. 1 of *The Cambridge Edition of the Works of Jonathan Swift* (Cambridge University Press, 2010). All other references in this chapter are to this edition will be abbreviated *T* and cited parenthetically.
5. Vickers, "Swift and the Baconian Idol," 87, 117nn1–2. Vickers provides a meticulously supported and exhaustive account of Bacon's presence in Swift's work. He does not, however, consider that Swift's formal borrowings from Bacon have consequences for the history of the representation of thought in early fiction.
6. Vickers, "Swift and the Baconian Idol," 92.
7. Sophie Gee points out that although Swift adopts disordered voices to attack the style and philosophy of his Whig enemies, "this understanding . . . obscures the important point that Swift's 'mad' speakers are not merely impersonations adopted for ironic and satirical effects. They express, rather, his anguish about claiming to know the self or to understand one's own interiority." "'Such Opinions Cannot Cohere': Swift's Inwardness," *Republics of Letters: A Journal for the Study of Knowledge, Politics, and the Arts* 4, no. 1 (2014): 14.
8. Bacon, *Advancement of Learning*, 234–35.
9. It might be strange to think of Swift as a practitioner of psychological realism. And yet, as Michael McKeon has argued in relation to *Gulliver's Travels*, "by subverting empirical epistemology, Swift contributes, as fully as Defoe does by sponsoring it, to the growth of modern ideas of realism and the internalized spirituality of the aesthetic" (*Origins of the English Novel*, 352–53).
10. Jonathan Swift, *Thoughts on Various Subjects*, in *The Prose Works of Jonathan Swift*, ed. Herbert Davis, vol. 1, *A Tale of a Tub, With Other Early Works, 1696–1707* (Basil Blackwell, 1939), 241. As Herbert Davis mentions in his introduction to the volume, this first version of "Thoughts on Various Subjects," printed in the *Miscellanies* of 1711, was likely begun around the time Swift was working on the *Tale* and "may even be regarded as little scraps left over from his work on *A Tale of a Tub*" (xxxv).
11. This topic has been addressed by Pat Rogers, who is especially helpful in correcting critical misunderstandings regarding Swift's relationship to

authority. Even norm-less satires, Rogers argues, can be in the service of orthodoxy. "Swift and the Idea of Authority," in *The World of Jonathan Swift*, ed. Brian Vickers (Basil Blackwell, 1968), 25–37, especially 34–35.

12. Scholars who focus on the moral vision of Swift's satire agree on his Anglican piety (though accounts of his orthodoxy vary) but disagree about his views regarding reason's role in a virtuous life. Irvin Ehrenpreis, for example, finds the Swift of the *Tale* a proponent of the Christian humanist idea that a gentlemanly education in virtue will ensure rational living. See Ehrenpreis, *Swift: The Man, His Works, and the Age*, vol. 1, *Mr Swift and His Contemporaries* (Methuen, 1962), 216–25. Claude Rawson, by contrast, focuses on Swift's anxiety concerning rationality. Rawson's discussion draws on the Augustan concepts of Nature and Reason. See Rawson, *Gulliver and the Gentle Reader: Studies in Swift and Our Time* (Routledge & Kegan Paul, 1973), 18–32.

13. For Swift as a contributor to novelistic innovation, see J. A. Downie, "Swift and the Making of the English Novel," in *Reading Swift: Papers from the Third Münster Symposium on Jonathan Swift*, ed. Hermann J. Real and Helgard Stöver-Leidig (Wilhelm Fink Verlag, 1998), 179–87; and Jenny Davidson, "Austen's Voices," in *Swift's Travels: Eighteenth-Century British Satire and Its Legacy*, ed. Nicholas Hudson and Aaron Santesso (Cambridge University Press, 2008), 233–50.

14. Watt, *Rise of the Novel*, 193.

15. For recent work that connects eighteenth-century empiricism to contemporary theories of mind via the early novel, see Jonathan Kramnick, "Empiricism, Cognitive Science, and the Novel," *Eighteenth Century* 48, no. 3 (2007): 263–85; and Kramnick, *Actions and Objects from Hobbes to Richardson* (Stanford University Press, 2010). For an insightful and comprehensive summary of recent scholarly responses to Watt's individualist thesis, see the opening of Jonathan Lamb, "'Lay Aside My Character': The Personate Novel and Beyond," *Eighteenth Century* 52, nos. 3–4 (2011): 271–74. Lamb draws on many of Watt's interlocutors from the past decades: Catherine Gallagher, *Nobody's Story: The Vanishing Acts of Women Writers in the Marketplace, 1670–1820* (University of California Press, 1994); Gallagher, "The Rise of Fictionality," in *The Novel*, ed. Franco Moretti (Princeton University Press, 2006), 1:336–63; Deidre Lynch, *The Economy of Character: Novels, Market Culture, and the Business of Inner Meaning* (University of Chicago Press, 1998); McKeon, *Secret History of Domesticity*; and McKeon, *Origins of the English Novel*.

16. Kramnick, "Empiricism, Cognitive Science, and the Novel," 273.

17. Sean Silver, however, argues that even the Enlightenment "plain style," as much as it eschewed rhetorical ornament, relied on metaphor to bind word to thing as closely as possible. Our embodied experience guides us

to the ideas and words we use for acts of intellection. Words relying on metaphor link body to mind, or—as Silver argues—they have been seen to connect two domains, "embodied action and conceptual understanding," that have never actually been that distinct. Silver, *The Mind Is a Collection: Case Studies in Eighteenth-Century Thought* (University of Pennsylvania Press, 2015), 10–11. In *The Experimental Imagination,* Tita Chico expands this discussion of metaphor within the plain style to "literariness," or figurative language more broadly, arguing that it was central not only to "the textual rendering of early scientific knowledge," but also to the conception of the objects of science and character of the scientist (18).

18. Bacon, *New Organon,* 28.
19. Bacon, *Advancement,* 234.
20. In emphasizing the innovative nature of Baconian aphoristic writing, I depart somewhat from traditional accounts, including the extended account of Baconian aphorizing in Brian Vickers, *Francis Bacon and Renaissance Prose* (Cambridge University Press, 1968), 60–95. Vickers explains that "despite Bacon's use of the aphorism as an ideal form against which to compare the restrictions of other men's systems," in the *Novum organum* he is in essence, "constructing a [new scientific] system [of induction] using an 'anti-systematic' form" (82).
21. Bacon, *New Organon,* 29–30.
22. According to Terry Castle, the instability of the *Tale*'s structure arises from Swift's attunement to the "epistemological complexities that afflict any reader's relations to a text": "No text is privileged in regard to truth. . . . Given its inescapably material status, every writing is a site for corruption, no matter what authority—natural, divine, or archetypal—we may wishfully invest in it." "Why the Houyhnhnms Don't Write: Swift, Satire, and the Fear of the Text," *Essays in Literature* 7 (1980): 35, 37.
23. In *Letter to a Young Gentleman,* Swift touches on the topic of "Commonplace Books," ultimately concluding "that Men of tolerable Intellectuals [should] rather trust to their own natural Reason, improved by a general Conversation with Books." *Irish Tracts, 1720–1723, and Sermons,* vol. 9 of *The Prose Works of Jonathan Swift,* ed. Herbert Davis (Blackwell, 1948), 75–76. For more on Swift and commonplacing, see Hugh Ormsby-Lennon, "Commonplace Swift," in *Reading Swift: Papers from the Third Münster Symposium on Jonathan Swift,* ed. Hermann J. Real and Helgard Stöver-Leidig (W. Fink, 1998), 13–44.
24. See Marcus Walsh, "Swift's *Tale of a Tub* and the mock book," in *Jonathan Swift and the Eighteenth-Century Book,* ed. Paddy Bullard and James McLaverty (Cambridge University Press, 2013): "The development of old and the appearance of new methods and forms of referencing were closely

associated with the development of a professional historiography and philology at the beginning of the long eighteenth century. Marginalia and footnotes were important elements among a large set of newly prominent scholarly apparatuses: contents lists, catalogues, commentaries, bibliographies, glossaries, indexes, all of them list-like, divisible, more or less Ramist" (103).

25. Erin Mackie persuasively presents Swift's reanimation of lifeless copies as an attack on the "modern" faith in transparent mediation in "Swift and Mimetic Sickness," *Eighteenth Century* 54, no. 3 (2013): 359–73. See also Castle, "Why the Houyhnhnms Don't Write," which follows a similar line of argument. Castle contends that Swift viewed the material, written Text (as opposed to speech) as inherently corrupt.
26. Bacon, *New Organon*, 34.
27. Ronald Paulson, *Theme and Structure in Swift's* Tale of a Tub (Yale University Press, 1960), 162.
28. Frances Ferguson, "Jane Austen, *Emma*, and the Impact of Form," *Modern Language Quarterly* 61, no. 1 (2000): 163–64.
29. See John Dussinger, *The Discourse of the Mind in Eighteenth-Century Fiction* (Mouton, 1974); and Joe Bray, *The Epistolary Novel: Representations of Consciousness* (Routledge, 2003).
30. This interest in the personal struggles of interiority is more visible in Bacon's chosen forms (aphorism, essay) than in the content of his work. His comments in the *Advancement* and *Novum organum* do, however, suggest that he believed the mind was always in motion. When deprived of matter (or, as Locke would say, objects of experience), the mind would, Bacon believed, "work upon itself, as the spider worketh his web," endlessly. *Advancement*, 140.
31. Morris Croll, "The Baroque Style in Prose," in *Seventeenth-Century Prose: Modern Essays in Criticism*, ed. Stanley Fish (Oxford University Press, 1971), 29.
32. Watt, *Rise of the Novel*, 175.
33. Jürgen Habermas, *The Structural Transformation of the Public Sphere: An Inquiry into a Category of Bourgeois Society*, trans. Thomas Burger and Frederick Lawrence (MIT Press, 1991), 28.
34. The Habermasian thesis has been critiqued from a wide variety of angles. For early and influential feminist critiques of Habermas, see Nancy Fraser, "What's Critical About Critical Theory?: The Case of Habermas and Gender," in *Feminism as Critique: Essays on the Politics of Gender in Late-Capitalist Societies*, ed. Seyla Benhabib and Drucilla Cornell (Polity, 1987), 31–56; and Fraser, "Rethinking the Public Sphere: A Contribution to the Critique of Actually Existing Democracy," in *The Phantom Public Sphere*,

ed. Bruce Robbins (University of Minnesota Press, 1993). See also Joan B. Landes, *Women and the Public Sphere in the Age of the French Revolution* (Cornell University Press, 1988).

35. See, for example, Patricia Meyer Spacks, *Privacy: Concealing the Eighteenth-Century Self* (University of Chicago Press, 2003); and Jenny Davidson, *Hypocrisy and the Politics of Politeness: Manners and Morals from Locke to Austen* (Cambridge University Press, 2004).
36. Bacon, *New Organon*, 36, 44.
37. Bacon, 25.
38. I do not mean to suggest that in the eighteenth century novelistic fiction was a stable "discursive category." Michael Gavin provides a useful summary of the "rise of fiction" debates beginning in the 1980s that speak to fictional diversity in the eighteenth century. One thing scholars such as Lennard Davis, Catherine Gallagher, and Mary Poovey agree on is that instability was the norm. See Gavin, "Real Robinson Crusoe," *Eighteenth-Century Fiction* 25, no. 2 (2012–13): 303–5. See also Srinivas Aravamudan, *Enlightenment Orientalism: Resisting the Rise of the Novel* (University of Chicago Press, 2011); and Paige, *Before Fiction*.
39. Hadot, *Philosophy as a Way of Life*, 81–89.
40. Hadot, 83.
41. Swift, *Thoughts on Various Subjects*, 244.
42. Among critics there is little consensus about the social, moral, or political utility of this inwardness effect. Angus Fletcher and Mike Benveniste put the debate this way: it regards whether free indirect discourse is "normative, even coercive" or "a source of empathy and social concern"; see "A Scientific Justification for Literature: Jane Austen's Free Indirect Style as Ethical Tool," *Journal of Narrative Theory* 43, no. 1 (Winter 2013): 8. For further discussions of the origins of the style and its operation in Austen, see Peter Bowen and Casey Finch, "'The Tittle-Tattle of Highbury': Gossip and Free Indirect Style in *Emma*," *Representations* 31 (1990): 1–18; Frances Ferguson's rejoinder, "Jane Austen, *Emma*, and the Impact of Form," *Modern Language Quarterly* 61 (2000): 157–80; Rebecca Richardson, "Dramatizing Intimacy: Confessions and Free Indirect Discourse in *Sense and Sensibility*," *ELH* 81, no. 1 (2014): 225–44; and Daniel Gunn, "Free Indirect Discourse and Narrative Authority in *Emma*," *Narrative* 12, no. 1 (2004): 35–54.
43. Claude Rawson, *Satire and Sentiment 1660–1830: Stress Points in the English Augustan Tradition* (Yale University Press, 1994), 269.
44. Jane Austen, *Emma*, ed. George Justice (W. W. Norton, [1815] 2012), 49.
45. Ann Banfield, *Unspeakable Sentences: Narration and Representation in the Language of Fiction* (Routledge & Kegan Paul, 1982).

46. Michael McKeon, *Theory of the Novel: A Historical Approach*, ed. Michael McKeon (Johns Hopkins University Press, 2000), 485.
47. Davidson, "Austen's Voices," 235.
48. Swift experimented with maxims in texts other than those discussed in this chapter, yet his deployment of the form is most dynamic when more traditional extended commentary on the maxim is absent (as in the *Tale*) or inappropriate (as in *Gulliver's Travels*). This is why, for example, the unfinished pamphlet "Maxims Controlled in Ireland" falls a bit flat by Swiftian standards. Swift's method is to dispute political-economic clichés to expose the ongoing subjection of Ireland and its people. (The "controlled" in the title means to try against the facts.) See "Maxims Controlled in Ireland," in *The Prose Works of Jonathan Swift*, ed. Herbert Davis, vol. 12, *Irish Tracts, 1728–1733* (Basil Blackwell, 1955), 131–37.

3. The New Realism of Literary Generalization in Richardson's *Clarissa*

1. Samuel Richardson, *Clarissa, or, The History of a Young Lady*, ed. Angus Ross (Penguin, 1985), 985. This edition follows the text of the first edition published in 1747–48. All subsequent references to this edition of the novel will be abbreviated *C* and cited parenthetically.
2. The *Collection* was first included in the *Passages Restored* in 1751 and was later published as a standalone volume with maxims drawn from Richardson's two other novels in 1755. For a more complete publication history of "A Collection of . . . Moral and Instructive Sentiments," see John Dussinger's introduction in vol. 3 of *Samuel Richardson's Published Commentary on* Clarissa, *1747–65*, ed. Florian Stuber (Pickering & Chatto, 1998), vii–xlvi. The maxims in Richardson's fictions continue to produce scholarly interest in the literary-formal conflict between maxim and realist fiction. See, for example, Leah Price, *The Anthology and the Rise of the Novel: From Richardson to George Eliot* (Cambridge University Press, 2000); and Raff, "Quixotes, Precepts, and Galateas."
3. Denis Diderot, "Eulogy of Richardson," in *Diderot's Selected Writings*, ed. Lester Crocker, trans. Derek Coltman (Macmillan, [1761] 1966), 108–12, 108. In the same essay Diderot writes of Richardson's prose: "All that Montaigne, Charron, La Rochefoucauld, and Nicole have put into the form of maxims, Richardson has expressed in narrative. But though an intelligent man could reconstruct most of the moralists' maxims from a thoughtful perusal of Richardson's work, he could never make a single page of Richardson out of all the moral maxims ever written" (108). For a thorough working out of the importance of conduct literature to the relationship between Richardson

and Diderot, see Rita Goldberg, *Sex and Enlightenment: Women in Richardson and Diderot* (Cambridge University Press, 1984).
4. Samuel Johnson, "*Rambler* 71 (November 20, 1750)," in *The Rambler* (J. Payne, 1752), 3:2.
5. William Warner, "A General Model of Realist Fictional Narrative," *ELH* 87 (2020): 855–80, 859.
6. Stephen Greenblatt, *Renaissance Self-Fashioning: From More to Shakespeare* (University of Chicago Press, 1980), 207–8.
7. Greenblatt, *Renaissance Self-Fashioning*, 208.
8. Mary Wortley Montagu, "To Lady Bute. October 20 [1755]," in *Clarissa: The Eighteenth-Century Response, 1747–1804*, vol. 1, ed. Lois E. Bueler (AMS, 2010), 62.
9. On *Clarissa* as outlining the "atomization" of personhood for women under heterosexual sex, see Kathleen Lubey, "Sexual Remembrance in *Clarissa*," *Eighteenth-Century Fiction* 29, no. 2 (Winter 2016–17): 151–78. Eighteenth-century English rape law focuses on protections for white women to the exclusion of others.
10. It also, I think, following Bruno Latour, leaves room for considering how the early realist novel considers a diverse array of possible compositions of reality. For Latour, inanimate objecthood is a modern human invention, the product of a process of the objectification of nonhuman things. Making persons objects—*elevating* them to the status of things—is part of Latour's larger goal of instantiating a "political epistemology" in which, by following the associations of human and nonhuman agents, we come to gain a "prospect" on the possibly different, and therefore changeable, compositions of our "common world." See Bruno Latour, *Reassembling the Social: An Introduction to Actor-Network Theory* (Oxford University Press, 2005); and Latour, "An Attempt at a 'Compositionist Manifesto,'" *New Literary History* 41 (2010): 471–90.
11. Diderot, "Eulogy," 108–9.
12. Samuel Johnson, "*Rambler* 4," in *Samuel Johnson: The Major Works*, ed. Donald Greene (Oxford University Press, 1984), 176.
13. John Bender, "Enlightenment Fiction and the Scientific Hypothesis," *Representations* 61 (1998): 6–28, 9.
14. John Bender, "Novel Knowledge: Judgment, Experience, Experiment," in *Fictions of Knowledge: Fact, Evidence, Doubt*, ed. Yota Batsaki, Subha Mukherji, and Jan-Melissa Schramm (Palgrave Macmillan, 2012), 131–51, 139. For more on the role of inference in the eighteenth-century novel's production of empirical knowledge, see Maioli, *Empiricism*.
15. For additional examples of work that draws parallels between the "plain style" of empirical report and literary realism, see Simon Schaffer and

Steven Shapin, *Leviathan and the Air-Pump*; and, of course, Watt, *Rise of the Novel*.
16. Gallagher, "Rise of Fictionality," 344.
17. Thompson, *Fictional Matter*, 3.
18. Thompson, 3.
19. Thompson, 12.
20. Thompson, 6.
21. Thompson, 11. Thompson uses the term "form" "to signify not an ideal blueprint imposed *on* matter but modes of experimental and representational access *to* matter." Such "modes of experimental and representational access" *are* "modes of knowledge" that "insinuate perceptual understanding back into particles . . . to figure their capacity to effect sensory experience" (15).
22. Thompson, 16.
23. I do not mean to overstate my case regarding the division between maxim and realist fiction. Assumptions regarding common wisdom undergird the early novel's representation of cause and effect, and exemplary characters and other novelistic examples were understood to be vivid vehicles for precept. See Michael Sayeau, "Realism and the Novel," in *The Cambridge Companion to the Novel*, ed. Eric Bulson (Cambridge University Press, 2018), 91–103. According to Sayeau, "Novelistic realism, at its most basic level, is a matter of conformity to certain rules of thumb and unwritten manuals of conventional wisdom about the sorts of things that are likely or unlikely to happen in a given situation" (95).
24. Roger Maioli discusses how novels—made of words—were subject to the Baconian critique that natural philosophers had been misled by words and needed to turn to—and learn from—nature for the knowledge they sought (*Empiricism*, 7).
25. On Richardson's complicated relationship to the signs of gentlemanly education and his debt to Edward Bysshe's *Art of English Poetry*, see Darryl P. Domingo, "'Well Observed by the Poet': Elias Brand and Richardson's British Ancients," *Eighteenth-Century Fiction* 24, no. 4 (Summer 2012): 597–622, in particular 599; and Price, *Anthology and the Rise of the Novel*, 40–41. On Brand's pedantry, see also Howard D. Weinbrot, *Menippean Satire Reconsidered: From Antiquity to the Eighteenth Century* (Johns Hopkins University Press, 2005).
26. Bacon, *Advancement of Learning*, 140.
27. Bacon, 140.
28. The argument that the novel "tells of the adventure of interiority" comes from Georg Lukács, *The Theory of the Novel: A Historico-Philosophical Essay on the Forms of Great Epic Literature*, trans. Anna Bostock (MIT

Press, 1971), 88–89. Sandra Macpherson draws on Lukács in *Harm's Way*, 5. Macpherson reminds us that "In Lukács, the adventure of interiority, admittedly, is a *via dolorosa,* and the novel of romantic disillusionment continues to feel tragic from the way it subjects a hero of 'increased subjectivity' to the 'crushing, equalizing universality of fate'" (Lukács, 137; quoted in Macpherson, 6).

29. Latour, "Compositionist Manifesto," 476.
30. Ferguson, "Rape and the Rise," 99.
31. Ferguson, 98, emphasis mine.
32. Ferguson, 102.
33. Ferguson, 105–6.
34. Ferguson, 106.
35. Ferguson, 106. "Mimesis of distinction" is Ferguson's term.
36. Wendy Anne Lee, *Failures of Feeling: Insensibility and the Novel* (Stanford University Press, 2018), 7. She goes on: "To this end I argue that Clarissa's life as a transparent, urban rape survivor installs a trenchant self-critique and sweeping countermodel within a Richardsonian tradition."
37. Frances Ferguson, "Rape and the Rise," 105.
38. As recorded in Bowers and Richetti's abridged *Clarissa,* the first of these quotations comes from Samuel Garth's satirical professional poem *The Dispensary* (1699) and the second from John Dryden's *Absalom and Achitophel,* part 1, 43–44. *Clarissa, or, The History of a Young Lady (An Abridged Edition),* by Samuel Richardson, ed. Toni Bowers and John Richetti (Broadview, 2010), 483n1.
39. Richardson, *Clarissa (Abridged),* 483n1. For additional considerations of the representational quality of Paper X and the "mad letters" more broadly, see Ronald Paulson, *Emblem and Expression: Meaning in English Art in the Eighteenth Century* (Thames & Hudson, 1975), 51; Terry Castle, *Clarissa's Ciphers: Meaning and Disruption in Richardson's "Clarissa"* (Cornell University Press, 1982), 120; Amit Yahav-Brown, "Reasonableness and Domestic Fiction," *ELH* 73, no. 4 (Winter 2006): 805–30, 816, 817; Kathryn Steele, "*Clarissa*'s Silence," *Eighteenth-Century Fiction* 23, no. 1 (2010): 1–34, 23; Helen Thompson, "Secondary Qualities and Masculine Form in *Clarissa* and *Sir Charles Grandison*," *Eighteenth-Century Fiction* 24, no. 2 (Winter 2011–12): 195–226, 207; and Lubey, "Sexual Remembrance in *Clarissa,*" 174–75.
40. I draw my observation from two eighteenth-century women's commonplace books—one by Catherine Springett and one by Ann Bromfield—in the manuscript collection of the William Andrews Clark Memorial Library. These manuscript commonplace books are now available to the public through the digital archive Calisphere.

41. This point was inspired by Deidre Shauna Lynch's examination of the "rogue archivists" of the Romantic era manuscript "scrapbook" (*not* commonplace book) in "Paper Slips: Album, Archiving, Accident," *Studies in Romanticism* 57, no. 1 (2018): 87–114, 99.
42. William Warner, "Reality and the Novel: Latour and the Uses of Fiction," *Eighteenth Century* 57, no. 2 (2016): 267–79, 276.
43. For example, in *Styles of Meaning and Meanings of Style in Richardson's Clarissa* (McGill-Queen's University Press, 1999), Gordon D. Fulton explores the maxim in the context of a shift from communal to individual forms of communication: "If the failure of proverbs as an interactive strategy suggests how *Clarissa* can be [related] to a long-term transition from public to private orientation in life and literature and to a growing ideology of individualism, so too can non-proverbial generalization, the style of meaning whose generic form Richardson refers to variously as the moral or instructive sentiment, aphorism, maxim, caution, observation, and reflection" (52).
44. Jacob Sider Jost, *Prose Immortality, 1711–1819* (University of Virginia Press, 2015), 82.
45. William Congreve, *The Mourning Bride, A Tragedy*, 3rd ed. (Jacob Tonson, [1697] 1703), 2.
46. Congreve, *Mourning Bride*, 5.
47. Diderot, *Eulogy*, 109.
48. These compilations draw primarily on Job and Ecclesiasticus. See Tom Keymer, "Richardson's *Meditations*: Clarissa's *Clarissa*," in *Samuel Richardson: Tercentenary Essays*, ed. by Margaret Anne Doody and Peter Sabor (Cambridge University Press, 1989), 89–109, 89. Keymer sees Clarissa's act of writing the meditations as reparative, assuming that the original method of consolidating her experience through first-person narration no longer works.
49. Price, *Anthology and the Rise of the Novel*, 19.
50. Steele, "*Clarissa*'s Silence," 23.
51. Steele, 24.
52. In her study of how late seventeenth- and early eighteenth-century texts use "non-human viewpoints" to construct the human viewpoint, Lynn Festa reminds us that "the material artifacts or cultural 'works' that purportedly build a common human world discriminate" (10). Fictions of the human, fictions that draw on the nonhuman to construct humanity, "can both supply the grounds for the emancipatory reclamation of revolutionary rights and serve as an instrument to legitimate oppressive hierarchies" (10). Lynn Festa, *Fiction Without Humanity: Person, Animal, Thing in Early Enlightenment Literature and Culture* (University of Pennsylvania Press, 2019).

53. Sandra Macpherson, "A Little Formalism," *ELH* 82, no. 2 (2015): 399.
54. Calls (such as Latour's, for example) to "reify" the human, to treat humans as things, can equally perpetuate injustice, or recreate it in new forms. Lynn Festa outlines the potentially problematic dimensions of the "posthuman" turn. Who, she asks, has the privilege to disavow humanity? (*Fiction Without Humanity*, 19–20). Indeed, how can we sing the emancipatory potential of objecthood with regard to a period in which the category of personhood excluded so many: the enslaved, the nonwhite, the non-Christian, those with disabilities, the poor?

4. Austen's Lessons Not Worth Knowing

1. Jane Austen, *Northanger Abbey, Lady Susan, The Watsons, and Sanditon*, ed. James Kinsley and John Davie (Oxford University Press, 2003), 11, 12; 12–13, 18. All other references to *Northanger Abbey* will be abbreviated *NA* and cited parenthetically.
2. Claudia Johnson, introduction and notes to *Northanger Abbey, Lady Susan, The Watsons, and Sanditon,* by Jane Austen, ed. James Kinsley and John Davie (Oxford University Press, 2003), 361n18. See also Jan Fergus, *Jane Austen: A Literary Life* (Macmillan, 1991), 37. The earliest edition of Dyche's schoolbook available through Eighteenth Century Collections Online is the second, 1710 edition. There the heroic couplet appears: "Despair of Nothing that you wou'd attain: / Unweari'd Diligence your Point will gain" (126). Versions of Dyche's couplet appears again in *The Lady's Accomptant and Best Accomplisher* (Robinson & Roberts, 1771), which contains not only moral lessons but lessons in arithmetic, money, bills of fare, receipts, and the like, and in J. Girrard, *Practical Lectures on Education, Spiritual and Temporal; Extracted from the Most Eminent Authors on That Subject* (Andrew Brice, 1756), 198. The full quatrain appears there as follows: "Quickly lay hold on Time while in your Power: / Be careful well to husband every Hour. / Despair of Nothing which you would attain; / Unwearied Diligence your Point will Gain."
3. This is what Johnson suggests in her notes to the Oxford edition.
4. Jane Austen, *Pride and Prejudice,* ed. Vivien Jones (Penguin, 2014), 33. All other references to *Pride and Prejudice* will be abbreviated *P* and cited parenthetically.
5. Indeed, a phrase from the maxim—"unwearied diligence"—appears elsewhere in the early pages of *Northanger Abbey* in a passage describing Mrs. Allen's and Catherine's progress through a crowded ballroom: "But to her utter amazement she found that to proceed along the room was by no means the way to disengage themselves from the crowd; it seemed

rather to increase as they went on, whereas she had imagined that when once fairly within the door, they should easily find seats and be able to watch the dances with perfect convenience. But this was far from being the case, and though by unwearied diligence they gained even the top of the room, their situation was just the same.... Still they moved on— something better was yet in view; and by a continued exertion of strength and ingenuity they found themselves at last in the passage behind the highest bench" (11).

6. D. A. Miller, *Jane Austen, or The Secret of Style* (Princeton University Press, 2003), 23.
7. Nineteenth-century interiority, D. A. Miller contends, is a source of joy and torment for those who cultivate it (14). In a world that is made "consistently intelligible" through the socially agreed upon manufacture of clear signs—the "expressive air" of a person or "meaningful looks" of a face—"insignificance" is a "grave[] charge" indeed, and "under its force, the human toothpick case [Robert Ferrars from *Sense and Sensibility*] comes to embody, beyond an individual shallowness of socialization and subjectivity, the danger of their general flattening, even undoing" (*Jane Austen*, 14).
8. Hannah More, *Strictures on the Modern System of Female Education*, 2 vols. (T. Cadell, 1799), 11.7.161; quoted in Price, Anthology and the Rise of the Novel, 74.
9. Price, *Anthology and the Rise of the Novel*, 74.
10. Silver, *Mind Is a Collection*, 33.
11. In the early stages of this project, I spent considerable time combing through early eighteenth-century editions of such works in Early English Books Online and Eighteenth Century Collections Online. See William Penn, *Some Fruits of Solitude in Reflections and Maxims Relating to the Conduct of Human Life*, 6th ed. (Thomas Northcott, 1702); and Penn, *More Fruits of Solitude: Being the Second Part of Reflections and Maxims, Relating to the Conduct of Humane Life* (T. Sowle, 1702).
12. Such material is the stuff of the conduct books featured in Nancy Armstrong's influential *Desire and Domestic Fiction* (1987).
13. It is worth considering such arguments alongside this book's earlier explication of Bacon's call for using "dispersed directions" to encourage empirical learning through doing, a mode of hands-on natural philosophical instruction premised on the separation of natural from human laws.
14. Price puts her argument regarding the anthology succinctly in her introduction: "At once the voice of authority and a challenge to prevailing models of authorship, the anthology traces its ambiguity to the late eighteenth century, when an organicist theory of the text and a proprietary

understanding of authorship gathered force at the same moment as legal and educational changes lent compilers new power" (*Anthology and the Rise*, 3). Price incorporates and builds on the arguments of Barbara Benedict, *Making the Modern Reader: Cultural Mediation in Early Modern Anthologies* (Princeton University Press, 1996).
15. Price, 68.
16. Price, 74.
17. Richardson had modeled himself initially as a "reader" and not a "writer"— a reader of others' letters. But he later assigned the status of reader to women and writer to himself alone. Richardson is, according to Price, deeply ambivalent "about the competing demands of establishing authority and engaging readers—or of representing epistolary exchange and claiming literary property" (Price, 39).
18. For critics of popular advice-givers such as Chesterfield, the maxim's rhetorical history taints the guidance it mediates. See Jenny Davidson, *Hypocrisy and the Politics of Politeness: Manners and Morals from Locke to Austen* (Cambridge University Press, 2004).
19. George Savile, Marquis of Halifax, *New-Year's-Gift, or, Advice to a Daughter*, 7th ed. (Printed for D. Midwinter & T. Leigh, 1701), 149. In their editorial notes to the 2016 Norton edition, Donald Gray and Mary Favret state that David Shapard "suggests a book by the rhetorician Hugh Blair (1718–1800) or a conduct book by Hester Chapone (1727–1801)" as a source for Mary's extract in this passage (15n7).
20. See Mary Poovey, *The Proper Lady and the Woman Writer: Ideology as Style in the Works of Mary Wollstonecraft, Mary Shelley, and Jane Austen* (University of Chicago Press, 1984); Audrey Bilger, *Laughing Feminism: Subversive Comedy in Frances Burney, Maria Edgeworth, and Jane Austen* (Wayne State University Press, 1998). When Bilger examines how the critique of conduct literature happens in comic writing by women in this period, for example, she explicitly identifies the "maxims" of patriarchal ideology that become clear in these women's ridicule of male supremacy (119). Indeed, citing the character Lady Cecilia in Maria Edgeworth's *Helen*, Bilger demonstrates how female characters laughing at sexist men often do so by distilling their behavior into sexist maxims, demonstrating their intellectual abilities while lampooning male sexism (119).
21. Bilger, *Laughing Feminism*, 73.
22. Bilger, 74–75.
23. This "having it both ways" holds true for both Elizabeth Bennet and Jane Austen herself. For more on this, see Mary Poovey, "Ideological Contradictions and the Consolations of Form," in *The Proper Lady and the Woman Writer: Ideology as Style in the Works of Mary Wollstonecraft, Mary Shelley, and Jane Austen* (University of Chicago Press, 1984), 172–207.

24. Paul Dawson, "Fictional Minds and Female Sexuality: The Consciousness Scene from *Pamela* to *Lady Chatterley's Lover*," *ELH* 86, no. 1 (Spring 2019): 161–88, 163.
25. See Ryle, "Jane Austen and the Moralists"; and David Gallop, "Jane Austen and the Aristotelian Ethic," *Philosophy and Literature* 23, no. 1 (1999): 96–106.
26. Dorrit Cohn, *Transparent Minds: Narrative Modes for Presenting Consciousness in Fiction* (Princeton University Press, 1978), 102.
27. In *Open Secrets: The Literature of Uncounted Experience* (Stanford University Press, 2008) Anne-Lise François asks us to consider a heroine like Fanny Price, who through her presence in a third-person narrative that occasionally slips into free indirect discourse, is "reliev[ed] . . . from first-person assertions" (224). For Thomas Salem Manganaro in *Against Better Judgment: Irrational Action and Literary Invention in the Long Eighteenth Century* (University of Virginia Press, 2022), free indirect discourse creates logically impossible sentences that mirror the conflicts and self-deceptions of an individual mind.
28. François, *Open Secrets*, 225.
29. Catherine is unlike other eighteenth-century heroines. She is neither blessed with innate perfections that put her at odds with her social environment nor is she a fully quixotic misinterpreter of daily life. In their introduction to the Cambridge edition of *Northanger Abbey* that they edit, Barbara M. Benedict and Deirdre Le Faye make the point that Catherine "seems not to learn anything from conduct books." Following Alison G. Sulloway's 1989 *Jane Austen and the Province of Womanhood*, Benedict and Le Faye attribute Austen's decision here to the fact that such conduct books "were written by men who often betrayed condescending attitudes toward women." Jane Austen, *Northanger Abbey* [1817], in *The Cambridge Edition of the Works of Jane Austen*, ed. Barbara M. Benedict and Deirdre Le Faye (Cambridge University Press, 2006).
30. Austen to James Stanier Clarke, December 11, 1815, in *Jane Austen's Letters*, 4th ed., ed. Deirdre Le Faye (Oxford University Press, 2011), 319.
31. See Hershinow, *Born Yesterday*. We might compare Hershinow's conclusions to those of Anne-Lise François in *Open Secrets*. François argues against the imperative that experience be *worth* something. François considers feminine naivete or not-knowing as the condition of unknowingness, exploring that condition's relationship to the unknown, especially that which is unknown to the self. François tracks the "legacy of the countervailing command to do good only in secrecy, in ways unknown even to oneself: 'But when thou doest alms, let not thy left hand know what thy right hand doth./That thine alms may be in secret; and thy Father which seeth in secret himself shall reward thee openly' (Matthew 6:3–4)" (26).

She goes on: "This surprising vindication not simply of secrecy but of 'unknowingness' finds an echo in the ways in which Western culture has historically defined feminine virtue as not knowing (modesty) and not doing (chastity)" (26). In François's words, unknowingness rejects "the Greek imperative 'know thyself,'" commanding instead "heedlessness—'do not advertise; do not add up; above all, do not count—remember to forget.'" Unknowingness "not only momentarily suspends the usual identification of moral conscience and consciousness, but presents the aporia of a command to obliviousness, which to heed is to betray and to forget is to obey, and in this it has obvious affinities with the often similarly vexed reclaiming of passivity and non manifestation as the site of the ethical found in certain strains of post-Enlightenment thought" (27).

32. Aaron Hanlon, *Empirical Knowledge in the Eighteenth-Century Novel: Beyond Realism* (Cambridge University Press, 2022), 41. "Stories of quixotes almost always incorporate realist elements that grant quixotism some degree of empirical plausibility in the physical world of the quixote. This is instrumental to the eighteenth-century rise of the quixotic novel as a vehicle for satirizing large-scale problems, such as social customs and legal and political systems. It is also instrumental to the increasing tendency, over the course of the eighteenth century, for readers to see quixotes as heroes and heroines in failed societies rather than dunces or objects of satire" (39).
33. Hanlon, *Empirical Knowledge*, 41.
34. Hanlon, 49.
35. Hanlon, 49.
36. Claudia L. Johnson, *Jane Austen: Women, Politics, and the Novel* (University of Chicago Press, 1988), 32.
37. Johnson, *Jane Austen*, 32–33. Johnson references Ronald Paulson's phrase "sensitive young girl," from *Representations of Revolution*, 221.
38. D. A. Miller, "Austen's Attitude," *Yale Journal of Criticism* 8 (1995): 1–5, 4.
39. Miller, "Austen's Attitude," 4.
40. Miller, 4.
41. Miller, *Jane Austen*, 19.

Conclusion

1. On Austen and comfort, see Yoon Sun Lee, "Jane Austen, Whiteness, and the Phenomenology of Comfort," *Keats-Shelley Journal* 70 (2021): 111–17.
2. Rachel Cusk, *Outline* (Farrar, Straus & Giroux, 2014), 21. All further references to this work will be cited parenthetically.
3. Interview by Alexandra Schwartz, *The New Yorker*, November 18, 2018, https://www.newyorker.com/culture/the-new-yorker-interview/i-dont-think-character-exists-anymore-a-conversation-with-rachel-cusk.

4. Schwartz, interview.
5. Schwartz, interview.
6. There are few theories of the maxim. Andrew Hui writes that aphorisms come before, against, and after philosophy. They are philosophy's ancestors and antagonists. "Central to my theory, then," he writes, "is that the aphorism is a dialectical play between fragments and systems." Hui, *A Theory of the Aphorism, from Confucius to Twitter* (Princeton University Press, 2019), 12.
7. Bender, *Ends of Enlightenment*, 26.
8. Culler makes this point in the process of defending an account of genre that is both "empirical" and "theoretical" and that responds to Ralph Cohen's account of genres "as open [historically contingent] systems" (65). That is, Culler argues "that conceptions of genres are not just accounts of what people of a particular period thought." He goes on: "It is crucial to the notion of genre as model that people might have been wrong about them, unaware of affinities or ignoring continuities in favor of more striking novelties, or recognizing only an attenuated version of a larger tradition. Genre study cannot be just a matter, for instance, of looking at what Renaissance critics say about genres and using only those categories for thinking about Renaissance literature, though of course one should try them out, while keeping in mind the possibility that more capacious and historically informed categories may be essential to grasping the full import and deepest resources of literary productions" (65). Jonathan Culler, "Lyric, History, and Genre" (2009), in *The Lyric Theory Reader: A Critical Anthology*, ed. Virginia Jackson and Yopie Prins (Johns Hopkins University Press, 2014), 63–77. See also Ralph Cohen, "History and Genre," *NLH* 17, no. 2 (1986): 203.
9. Austen, *Northanger Abbey*, 81.
10. Scholars have contested and continue to contest the claim that the novel was "consolidated" in Britain in the 1740s. Since I end with Austen, I am interested in the claim and revisions to it. See Srinivas Aravamudan's *Enlightenment Orientalism*.
11. There are a range of reasons for being disillusioned—or even disgusted—with the novel's obsessive attention to the inner life. John Bender's account of the eighteenth-century novel and D. A. Miller's account of the nineteenth-century novel both begin from Foucauldian premises of a late-Enlightenment cultural shift toward the embedding of discursive forces in private life.

BIBLIOGRAPHY

Aravamudan, Srinivas. *Enlightenment Orientalism: Resisting the Rise of the Novel*. University of Chicago Press, 2011.
Austen, Jane. *Emma* [1815]. Edited by George Justice. W. W. Norton, 2012.
———. *Northanger Abbey, Lady Susan, The Watsons, and Sanditon*. Edited by James Kinsley and John Davie. Oxford University Press, 2003.
———. *Pride and Prejudice* [1813]. Edited by Vivien Jones. Penguin, 2014.
Bacon, Francis. *The Advancement of Learning* [1605]. In *The Major Works*. Edited by Brian Vickers. Oxford University Press, 2008.
———. *Of the Advancement and Proficiencie of Learning: or the Partitions of Sciences*. Translated by Gilbert Wats. Thomas Williams, 1674.
———. *The New Organon* [1620]. Edited by Lisa Jardine and Michael Silverthorne. Translated by Michael Silverthorne. Cambridge University Press, 2000.
———. *The Philosophical Works of Francis Bacon . . . Methodized, and Made English, from the Originals, with Occasional Notes, to Explain What Is Obscure*. Edited and translated by Peter Shaw. J. J. & P. Knapton, D. Midwinter & A. Ward, A. Bettesworth & C. Hitch, J. Pemberton, J. Osborn & T. Longman, C. Rivington, F. Clay, J. Batley, R. Hett, & T. Hatchett, 1733.
Ballaster, Ros. *Seductive Forms: Women's Amatory Fiction from 1684 to 1740*. Clarendon, 1992.
Banfield, Ann. *Unspeakable Sentences: Narration and Representation in the Language of Fiction*. Routledge & Kegan Paul, 1982.
Barthes, Roland. "La Rochefoucauld: 'Reflections or Sentences and Maxims.'" In *New Critical Essays*, 3–22. Translated by Richard Howard. Hill & Wang, 1980.
Behn, Aphra. "The Fair Jilt; or, The Amours of Prince *Tarquin* and *Miranda*." In *All the Histories and Novels Written by the Late Ingenious Mrs. Behn*, 142–200. R. Wellington, 1705.
———. *The Lover's Watch*. In *The Works of Aphra Behn*, vol. 6. Edited by Montague Summers. Benjamin Blom, [1915] 1967.
———. *Seneca Unmasqued: A Bilingual Edition of Aphra Behn's Translation of La Rochefoucauld's Maximes*. Edited by Irwin Primer. AMS, 2001.
Bender, John. *Ends of Enlightenment*. Stanford University Press, 2012.
———. "Enlightenment Fiction and the Scientific Hypothesis." *Representations* 61 (1998): 6–28.

———. "Novel Knowledge: Judgment, Experience, Experiment." In *Fictions of Knowledge: Fact, Evidence, Doubt*, edited by Yota Batsaki, Subha Mukherji, and Jan-Melissa Schramm, 131–51. Palgrave Macmillan, 2012.

Bilger, Audrey. *Laughing Feminism: Subversive Comedy in Frances Burney, Maria Edgeworth, and Jane Austen*. Wayne State University Press, 1998.

Blackmore, E. H., and A. M., and Francine Giguère. Introduction to *Collected Maxims and Other Reflections* by François de La Rochefoucauld, ix–xxxii. Edited and translated by E. H. and A. M. Blackmore and Francine Giguère. Oxford University Press, 2007.

Brooke, Christopher. *Philosophic Pride: Stoicism and Political Thought from Lipsius to Rousseau*. Princeton University Press, 2012.

Carter, Elizabeth. Introduction to *All the Works of Epictetus, Which Are Now Extant; Consisting of His Discourses . . . In Four Books* by Epictetus, ii–xxxiv. Translated by Elizabeth Carter. S. Richardson, 1758.

Castle, Terry. *Clarissa's Ciphers: Meaning and Disruption in Richardson's Clarissa*. Cornell University Press, 1982.

———. "Why the Houyhnhnms Don't Write: Swift, Satire, and the Fear of the Text." *Essays in Literature* 7 (1980): 31–44.

Cave, Terence. Introduction to *La Princesse de Clèves*. Translated by Terence Cave, vii–xxx. Oxford University Press, 1992

Cervantes Saavedra, Miguel de. *The Ingenious Hidalgo Don Quixote de la Mancha*. Translated by John Rutherford. Penguin, 2000.

Chambers, Ephraim. *Cyclopædia: or, an Universal Dictionary of Arts and Sciences*. Vol. 1. James & John Knapton, John Darby, Daniel Midwinter, Arthur Bettesworth, John Senex, Robert Gosling, John Pemberton, William & John Innys, John Osborn & Thomas Longman, Charles Rivington, John Hooke, Ranew Robinson, Francis Clay, Aaron Ward, Edward Symon, Daniel Browne, Andrew Johnson, & Thomas Osborn, 1728.

Chico, Tita. *The Experimental Imagination: Literary Knowledge and Science in the British Enlightenment*. Stanford University Press, 2018.

Cohn, Dorrit. *Transparent Minds: Narrative Modes for Presenting Consciousness in Fiction*. Princeton University Press, 1978.

Congreve, William. *The Mourning Bride, A Tragedy*. 3rd ed. N.p., 1703.

Croll, Morris. "The Baroque Style in Prose." In *Seventeenth-Century Prose: Modern Essays in Criticism*, edited by Stanley Fish, 26–52. Oxford University Press, 1971.

Cusk, Rachel. *Outline*. Farrar, Straus & Giroux, 2014.

Davidson, Jenny. "Austen's Voices." In *Swift's Travels: Eighteenth-Century British Satire and Its Legacy*, edited by Nicholas Hudson and Aaron Santesso, 233–50. Cambridge University Press, 2008.

———. *Hypocrisy and the Politics of Politeness: Manners and Morals from Locke to Austen*. Cambridge University Press, 2004.

Dawson, Paul. "Fictional Minds and Female Sexuality: The Consciousness Scene from *Pamela* to *Lady Chatterley's Lover*." *ELH* 86, no. 1 (Spring 2019): 161–88.

Defoe, Daniel. *Robinson Crusoe*. Edited by Thomas Keymer. Oxford University Press, 2007.

Diderot, Denis. "Eulogy of Richardson" [1761]. In *Diderot's Selected Writings*, edited by Lester Crocker, translated by Derek Coltman, 108–12. Macmillan, 1966.

Dienstag, Joshua Foa. *Pessimism: Philosophy, Ethic, Spirit*. Princeton University Press, 2006.

Domingo, Darryl P. "'Well Observed by the Poet': Elias Brand and Richardson's British Ancients." *Eighteenth-Century Fiction* 24, no. 4 (Summer 2012): 597–622.

Downie, J. A. "Swift and the Making of the English Novel." In *Reading Swift: Papers from the Third Münster Symposium on Jonathan Swift*, edited by Hermann J. Real and Helgard Stöver-Leidig, 179–87. Wilhelm Fink Verlag, 1998.

Dussinger, John. Introduction to *A Collection of the Moral and Instructive Sentiments, Maxims, Cautions, and Reflections, Contained in the Histories of Pamela, Clarissa, and Sir Charles Grandison* [1755]. Vol. 3 of *Samuel Richardson's Published Commentary on* Clarissa, *1747–65*, edited by Florian Stuber, vii–xlvi. Pickering & Chatto, 1998.

Ehrenpreis, Irvin. *Mr Swift and His Contemporaries*. Vol. 1 of *Swift: The Man, His Works, and the Age*. Edited by Irvin Ehrenpreis. Methuen, 1962.

Ferguson, Frances. "Jane Austen, *Emma*, and the Impact of Form." *Modern Language Quarterly* 61, no. 1 (2000): 157–80.

———. "Rape and the Rise of the Novel." *Representations* 20 (Autumn 1987): 88–112.

Festa, Lynn. *Fiction Without Humanity: Person, Animal, Thing in Early Enlightenment Literature and Culture*. University of Pennsylvania Press, 2019.

Finch, Casey, and Peter Bowen. "'The Tittle-Tattle of Highbury': Gossip and Free Indirect Style in *Emma*." *Representations* 31 (1990): 1–18.

Fletcher, Angus, and Mike Benveniste. "A Scientific Justification for Literature: Jane Austen's Free Indirect Style as Ethical Tool." *Journal of Narrative Theory* 43, no. 1 (Winter 2013): 1–18.

Force, Pierre. *Self-Interest Before Adam Smith: A Genealogy of Economic Science*. Cambridge University Press, 2003.

François, Anne-Lise. *Open Secrets: The Literature of Uncounted Experience*. Stanford University Press, 2008.

Fulton, Gordon D. *Styles of Meaning and Meanings of Style in Richardson's* Clarissa. McGill-Queen's University Press, 1999.

Gallagher, Catherine. "The Rise of Fictionality." In *The Novel*, edited by Franco Moretti, 1:336–63. Princeton University Press, 2006.

Gee, Sophie. "'Such Opinions Cannot Cohere': Swift's Inwardness." *Republics of Letters: A Journal for the Study of Knowledge, Politics, and the Arts* 4, no. 1 (2014). http://arcade.stanford.edu/rofl/such-opinions-cannot-cohere-swifts-inwardness.

Girten, Kristin M. "Mingling with Matter: Tactile Microscopy and the Philosophic Mind in Brobdingnag and Beyond." *Eighteenth Century* 54, no. 4 (Winter 2013): 497–520.

———. *Sensitive Witnesses: Feminist Materialism in the British Enlightenment.* Stanford University Press, 2024.

Greenblatt, Stephen. *Renaissance Self-Fashioning: From More to Shakespeare.* University of Chicago Press, 1980.

Gunn, Daniel. "Free Indirect Discourse and Narrative Authority in *Emma*." *Narrative* 12, no. 1 (2004): 35–54.

Habermas, Jürgen. *The Structural Transformation of the Public Sphere: An Inquiry into a Category of Bourgeois Society.* Translated by Thomas Burger and Frederick Lawrence. MIT Press, 1991.

Hadot, Pierre. *Philosophy as a Way of Life: Spiritual Exercises from Socrates to Foucault.* Edited by Arnold I. Davidson. Translated by Michael Chase. Blackwell, 1995.

Hanlon, Aaron. *Empirical Knowledge in the Eighteenth-Century Novel: Beyond Realism.* Cambridge University Press, 2022.

Hershinow, Stephanie Insley. *Born Yesterday: Inexperience and the Early Realist Novel.* Johns Hopkins University Press, 2019.

Hui, Andrew. *A Theory of the Aphorism, from Confucius to Twitter.* Princeton University Press, 2019.

Jardine, Lisa. *Francis Bacon: Discovery and the Art of Discourse.* Cambridge University Press, 1974.

Johnson, Claudia L. Introduction and notes to *Northanger Abbey, Lady Susan, The Watsons, and Sanditon,* by Jane Austen, vii–xxxvii. Edited by James Kinsley and John Davie. Oxford University Press, 2003.

———. *Jane Austen: Women, Politics, and the Novel.* University of Chicago Press, 1988.

Johnson, Samuel. *A Dictionary of the English Language.* 2 vols. J. & P. Knaptor, T. & T. Longman, C. Hitch & L. Hawes, A. Millar, and R. & J. Dodsley, 1755.

———. *The Rambler.* Vol. 3. J. Payne, 1752.

———. "Rambler 4." In *Samuel Johnson: The Major Works,* edited by Donald Greene, 174–79. Oxford University Press, 1984.

Jost, Jacob Sider. *Prose Immortality, 1711–1819.* University of Virginia Press, 2015.

Kareem, Sarah Tindal. *Eighteenth-Century Fiction and the Reinvention of Wonder.* Oxford University Press, 2014.

Keymer, Thomas. Introduction to *Robinson Crusoe,* by Daniel Defoe, edited by Thomas Keymer, vii–xxxix. Oxford University Press, 2007.

———. "Richardson's *Meditations*: Clarissa's *Clarissa*." In *Samuel Richardson: Tercentenary Essays*, edited by Margaret Anne Doody and Peter Sabor, 89–109. Cambridge University Press, 1989.
Kramnick, Jonathan. *Actions and Objects from Hobbes to Richardson*. Stanford University Press, 2010.
———. "Empiricism, Cognitive Science, and the Novel." *Eighteenth Century* 48, no. 3 (2007): 263–85.
La Rochefoucauld, François de. *Collected Maxims and Other Reflections*. Edited and translated by E. H. and A. M. Blackmore and Francine Giguère. Oxford University Press, 2007.
Lamb, Jonathan. "'Lay Aside My Character': The Personate Novel and Beyond." *Eighteenth Century* 52, nos. 3–4 (2011): 271–87.
Lafayette, Madame de. *La Princesse de Clèves*. Translated by Terence Cave. Oxford University Press, 1992.
Latour, Bruno. "An Attempt at a 'Compositionist Manifesto.'" *New Literary History* 41 (2010): 471–90.
———. *Reassembling the Social: An Introduction to Actor-Network Theory*. Oxford University Press, 2005.
Lee, Wendy Anne. *Failures of Feeling: Insensibility and the Novel*. Stanford University Press, 2018.
Lee, Yoon Sun. "Jane Austen, Whiteness, and the Phenomenology of Comfort." *Keats–Shelley Journal* 70 (2021): 111–17.
Lewis, Rhodri. "Francis Bacon, Allegory, and the Uses of Myth." *Review of English Studies* 61, no. 250 (2010): 360–89.
———. "A Kind of Sagacity: Francis Bacon, the *Ars memoriae*, and the Pursuit of Natural Knowledge." *Intellectual History Review* 19, no. 2 (2009): 155–75.
Locke, John. *An Essay Concerning Human Understanding* [1689]. Edited by Peter H. Nidditch. Clarendon, 1975.
Lubey, Kathleen. "Sexual Remembrance in *Clarissa*." *Eighteenth-Century Fiction* 29, no. 2 (Winter 2016–17): 151–78.
Lukács, Georg. *The Theory of the Novel: A Historico-Philosophical Essay on the Forms of Great Epic Literature*. Translated by Anna Bostock. MIT Press, 1971.
Lynch, Deidre Shauna. *The Economy of Character: Novels, Market Culture, and the Business of Inner Meaning*. Chicago University Press, 1998.
———. "Paper Slips: Album, Archiving, Accident." *Studies in Romanticism* 57, no. 1 (2018): 87–114.
Mackie, Erin. "Swift and Mimetic Sickness." *Eighteenth Century* 54, no. 3 (2013): 359–73.
Macpherson, Sandra. *Harm's Way: Tragic Responsibility and the Novel Form*. Johns Hopkins University Press, 2010.
Maioli, Roger. *Empiricism and the Early Theory of the Novel: Fielding to Austen*. Palgrave Macmillan, 2016.

Manganaro, Thomas Salem. *Against Better Judgment: Irrational Action and Literary Invention in the Long Eighteenth Century.* University of Virginia Press, 2022.

Mannheimer, Katherine. "Celestial Bodies: Readerly Rapture as Theatrical Spectacle in Aphra Behn's *Emperor of the Moon*." *Restoration* 35, no. 1 (2011): 39–60.

McKeon, Michael. *The Origins of the English Novel, 1600–1740.* Johns Hopkins University Press, [1987] 2002.

———. *The Secret History of Domesticity: Public, Private, and the Division of Knowledge.* Johns Hopkins University Press, 2005.

———. *Theory of the Novel: A Historical Approach.* Edited by Michael McKeon. Johns Hopkins University Press, 2000.

Miller, D. A. "Austen's Attitude." *Yale Journal of Criticism* 8 (1995): 1–5.

———. *Jane Austen, or The Secret of Style.* Princeton University Press, 2003.

Miscellany, Being a Collection of Poems by Several Hands. J. Hindmarsh, 1685.

Montagu, Mary Wortley. "To Lady Bute. October 20 [1755]." In Vol. 1 of *Clarissa: The Eighteenth-Century Response, 1747–1804*, edited by Lois E. Bueler, 60–62. AMS, 2010.

Newman, Karen. "'Wit's Great Columbus': Aphra Behn Translates *La Montre*, or *The Lover's Watch*." *Shakespeare Studies* 48 (2020): 101–7.

Nightingale, Andrea. "Broken Knowledge." In *The Re-Enchantment of the World: Secular Magic in a Rational Age*, edited by Joshua Landy and Michael Saler, 15–37. Stanford University Press, 2009.

Ormsby-Lennon, Hugh. "Commonplace Swift." In *Reading Swift: Papers from the Third Münster Symposium on Jonathan Swift*, edited by Hermann J. Real and Helgard Stöver-Leidig, 13–44. W. Fink, 1998.

Paige, Nicholas. *Before Fiction: The Ancien Régime of the Novel.* University of Pennsylvania Press, 2011.

Pasanek, Brad. *Metaphors of Mind: An Eighteenth-Century Dictionary.* Johns Hopkins University Press, 2015.

Paulson, Ronald. *Emblem and Expression: Meaning in English Art in the Eighteenth Century.* Thames & Hudson, 1975.

———. *Theme and Structure in Swift's* Tale of a Tub. Yale University Press, 1960.

Poovey, Mary. *The Proper Lady and the Woman Writer: Ideology as Style in the Works of Mary Wollstonecraft, Mary Shelley, and Jane Austen.* University of Chicago Press, 1984.

Price, Leah. *The Anthology and the Rise of the Novel: From Richardson to George Eliot.* Cambridge University Press, 2000.

Primer, Irwin. Introduction to *Seneca Unmasqued: A Bilingual Edition of Aphra Behn's Translation of La Rochefoucauld's Maxims*, edited by Irwin Primer, vii–xliii. AMS, 2001.

Proctor, Robert N. "Agnotology: A Missing Term to Describe the Cultural Production of Ignorance (and Its Study)." In *Agnotology: The Making and Unmaking of Ignorance,* edited by Robert N. Proctor and Londa Schiebinger, 1–33. Stanford University Press, 2008.
Proctor, Robert N., and Londa Schiebinger, eds. *Agnotology: The Making and Unmaking of Ignorance.* Stanford University Press, 2008.
Raff, Sarah. "Quixotes, Precepts, and Galateas: The Didactic Novel in Eighteenth-Century Britain." *Comparative Literature Studies* 43, no. 4 (2006): 466–81.
Rawson, Claude. *Gulliver and the Gentle Reader: Studies in Swift and our Time.* Routledge & Kegan Paul, 1973.
———. *Satire and Sentiment 1660–1830: Stress Points in the English Augustan Tradition.* Yale University Press, 1994.
Richardson, Rebecca. "Dramatizing Intimacy: Confessions and Free Indirect Discourse in *Sense and Sensibility.*" *ELH* 81, no. 1 (2014): 225–44.
Richardson, Samuel. *Clarissa, or, The History of a Young Lady* [1747–48]. Edited by Angus Ross. Penguin, 1985.
———. *A Collection of the Moral and Instructive Sentiments, Maxims, Cautions, and Reflexions, Contained in the Histories of Pamela, Clarissa, and Sir Charles Grandison.* Samuel Richardson, 1755.
Robert, Marthe. "From *Origins of the Novel.*" In *Theory of the Novel: A Historical Approach,* edited by Michael McKeon, 57–69. Johns Hopkins University Press, 2000.
Rogers, G. A. J. "The Intellectual Setting and Aims of the *Essay.*" In *The Cambridge Companion to Locke's "Essay Concerning Human Understanding,"* edited by Lex Newman, 7–32. Cambridge University Press, 2007.
Rogers, Pat. "Swift and the Idea of Authority." In *The World of Jonathan Swift,* edited by Brian Vickers, 25–37. Basil Blackwell, 1968.
Sayeau, Michael. "Realism and the Novel." In *The Cambridge Companion to the Novel,* edited by Eric Bulson, 91–103. Cambridge University Press, 2018.
Serjeantson, R. W. "Proof and Persuasion." In *Early Modern Science,* edited by Katherine Park and Lorraine Daston, 132–75. Vol. 3 of *The Cambridge History of Science.* Cambridge University Press, 2006.
Shapin, Steven, and Simon Schaffer. *Leviathan and the Air-Pump: Hobbes, Boyle, and the Experimental Life.* Princeton University Press, [1985] 2011.
Silver, Sean. *The Mind Is a Collection: Case Studies in Eighteenth-Century Thought.* University of Pennsylvania Press, 2015.
Snider, Alvin. "Atoms and Seeds: Aphra Behn's Lucretius." *CLIO* 33, no. 1 (2003): 1–24.
Staves, Susan. *A Literary History of Women's Writing in Britain, 1660–1789.* Cambridge University Press, 2006.

Steele, Kathryn. "*Clarissa*'s Silence." *Eighteenth-Century Fiction* 23, no. 1 (2010): 1–34.
Swift, Jonathan. *The Battel of the Books* [1704]. In *A Tale of a Tub and Other Works*, edited by Marcus Walsh, 137–64. Vol. 1 of *The Cambridge Edition of the Works of Jonathan Swift*. Cambridge University Press, 2010.

———. *Gulliver's Travels* [1726]. Edited by David Womersley. Vol. 16 of *The Cambridge Edition of the Works of Jonathan Swift*. Cambridge University Press, 2010.

———. "Maxims Controlled in Ireland." In *Irish Tracts, 1728–1733*, edited by Herbert Davis, 131–37. Vol. 12 of *The Prose Works of Jonathan Swift*. Basil Blackwell, 1955.

———. "Swift to Alexander Pope (Sep. 29, 1725)." In *The Writings of Jonathan Swift*, edited by Robert A. Greenberg and William Bowman Piper, 584–85. W. W. Norton, 1973.

———. "Swift to Alexander Pope (Nov. 26, 1725)." In *The Writings of Jonathan Swift*, edited by Robert A. Greenberg and William Bowman Piper, 585–86. W. W. Norton, 1973.

———. *A Tale of a Tub* [1704]. In *A Tale of a Tub and Other Works*, edited by Marcus Walsh. Vol 1 of *The Cambridge Edition of the Works of Jonathan Swift*. Cambridge University Press, 2010.

———. "*The Tatler*, No. 5 (1711)." In *The Writings of Jonathan Swift*, edited by Robert A. Greenberg and William Bowman Piper, 456–60. W. W. Norton, 1973.

———. *Thoughts on Various Subjects*. In *A Tale of a Tub, with Other Early Works, 1696–1707*, edited by Herbert Davis. Vol. 1 of *The Prose Works of Jonathan Swift*. Basil Blackwell, 1939.

Thompson, Helen. *Fictional Matter: Empiricism, Corpuscles, and the Novel*. University of Pennsylvania Press, 2016.

———. "Secondary Qualities and Masculine Form in *Clarissa* and *Sir Charles Grandison*." *Eighteenth-Century Fiction* 24, no. 2 (Winter 2011–12): 195–226.

Todd, Janet. *The Secret Life of Aphra Behn*. Rutgers University Press, 1997.

Tucker, Joseph E. "The Earliest English Translation of La Rochefoucauld's *Maxims*." *Modern Language Notes* 64, no. 6 (1949): 413–15.

Valenza, Robin. "How Literature Becomes Knowledge: A Case Study." *ELH* 76, no. 1 (2009): 215–45.

Vickers, Brian. *Francis Bacon and Renaissance Prose*. Cambridge University Press, 1968.

———. "Swift and the Baconian Idol." In *The World of Jonathan Swift*, edited by Brian Vickers, 87–128. Basil Blackwell, 1968.

Walsh, Marcus. "Swift's 'Tale of a Tub' and the mock book." In *Jonathan Swift and the Eighteenth-Century Book*, edited by Paddy Bullard and James McLaverty, 101–18. Cambridge University Press, 2013.

Warner, William. "A General Model of Realist Fictional Narrative." *ELH* 87 (2020): 855–80.

———. "Reality and the Novel: Latour and the Uses of Fiction." *Eighteenth Century* 57, no. 2 (2016): 267–79.

Watt, Ian. *The Rise of the Novel: Studies in Defoe, Richardson, and Fielding.* University of California Press, [1957] 2001.

Wehling, Peter. "Why Science Does Not Know: A Brief History of (the Notion of) Scientific Ignorance in the Twentieth and Early Twenty-First Centuries." *Journal for the History of Knowledge* 2, no. 1 (2021): 1–13.

Weinbrot, Howard D. *Menippean Satire Reconsidered: From Antiquity to the Eighteenth Century.* Johns Hopkins University Press, 2005.

Yahav-Brown, Amit. "Reasonableness and Domestic Fiction." *ELH* 73, no. 4 (Winter 2006): 805–30.

Zitin, Abigail. *Practical Form: Abstraction, Technique, and Beauty in Eighteenth-Century Aesthetics.* Yale University Press, 2020.

INDEX

amatory fiction, 38–41, 48, 54, 62, 85–86
apatheia (equanimity), 8–10
aphorisms, 1–2, 5–10, 167n6; empiricism and, 56–57, 69–71; as method, 20, 25–29, 31–33, 44–45, 56–58, 61–62, 65–77, 79, 106, 130, 141n64, 141n69, 154n20, 155n30; as opposed to maxims, 16. *See also* Bacon, Francis: aphorisms; maxims
Aristotle, 1–2, 18–19, 21–22, 27–29, 31–32, 44–45, 48–49, 58–59, 76, 92–93, 141n68
Armstrong, Nancy, 106–7
atomism, 55–56. *See also* materialism
Augustine, 44, 53–54, 147n19; *The City of God*, 46
Austen, Jane, 1–2, 10–11, 107–8, 132; *Emma*, 74, 80–81; juvenilia, 119–20; *Northanger Abbey*, 36, 103–6, 110–12, 116–20, 123, 125, 129–30, 165n29; *Pride and Prejudice*, 36, 104, 109–15, 120–23, 132; *Sense and Sensibility*, 121

Bacon, Francis, 10–11, 30–34, 37–38, 41, 55, 60–63, 82–83, 91–93, 128–30, 138n29, 140nn54–55, 142n82, 143n83, 149n51; *The Advancement of Learning*, 21–27, 29–30, 65–66, 140n56, 155n30; aphorisms, 5–10, 25–29, 31–33, 44–45, 56–58, 61–62, 65–77, 79, 106, 130, 141n64, 141n69, 154n20, 155n30; *De augmentis scientiarum*, 23–24, 65–66; *Essays*, 27, 65–66, 141n69; "The Great Renewal" (*Instauratio magna*), 5–7, 136n14; *Novum organum*, 1–2, 5–7, 18–19, 27–30, 56–57, 70–72, 76–77, 141nn68–69, 155n30
Ballaster, Ros, 54

Banfield, Ann, 81–82
Barthes, Roland, 51–52
Behn, Aphra, 1–2, 10–11, 37–38, 62–63, 85–86, 131–32, 149nn45–46, 149n51; *A Discovery of New Worlds*, 38–39; *The Emperor of the Moon*, 38–39; *The Fair Jilt*, 5–7, 38–39; *The History of the Nun*, 38–39; *Love-Letters Between a Nobleman and His Sister*, 38–39; *The Lover's Watch: or, The Art of Making Love*, 60–61; *Seneca Unmasqued*, 8–10, 34–35, 37–46, 49–62, 132–33, 137nn22–23, 144n1, 145n6, 146n9, 150nn65–66, 151n69; "To the Unknown Daphnis," 55–56. *See also* La Rochefoucauld, François de
Bender, John, 3–4, 18, 89–92, 167n11
Bentley, Richard, 71–73
bildungsroman, 117–19
Bilger, Audrey, 109–10, 164n20
Blackmore, A. M. and E. H., 50–51, 137n23
Bonnecorse, Balthazar de, *La Montre*, 60–61, 150nn65–66, 151n69
Boyle, Robert, 3–7, 19–20, 90–92

cannibalism, 2, 79–80
Carter, Elizabeth, 47–48
Cervantes Saavedra, Miguel de, *Don Quixote*, 3
Chambers, Ephraim, *Cyclopædia*, 16
Chambers, Ross, 54
chiasmus, 7–8
Chico, Tita, 3–4, 56–57, 135n6, 145n4, 153n17
class: gender and, 104–7, 116–17, 131; knowledge and, 92–93, 131
Cohn, Dorrit, 115
compositionism, 93–94

179

conduct books, 8–10, 106, 108–12, 144n1, 157n3, 163nn11–12, 164n20, 165n29
Congreve, William, *The Mourning Bride*, 98–102
Cowley, Abraham, 29–30
Creech, Thomas, 55–56, 149n45
Croll, Morris, 75
Culler, Jonathan, 129–30, 167n8
Cusk, Rachel, *Outline*, 125–28

Davidson, Jenny, 82
Dawson, Paul, 110–12
Defoe, Daniel, *Robinson Crusoe*, 1–5
Diderot, Denis, 86–87, 89–92, 100, 157n3
Dienstag, Joshua, 57–58
Dyche, Thomas, 104, 162n2

empiricism, 5–7, 12–13, 16–17, 19–20, 29–30, 32–33, 35, 37–38, 64–72, 75–76, 86–87, 89–93, 97–98, 102, 117–20, 128, 131, 143n83. *See also* science
Epictetus, 47–48
Epicureanism, 37, 55, 78–79, 146n15, 149nn45–46, 149n51
epigrammaticism, 58
epistolary fiction, 13–14, 60–61, 69–70, 74, 94–97, 164n17
Esprit, Jacques, 44–45
example, 78–80, 89–90, 103–4, 159n23
experience: embodiment and, 29, 37, 106, 121–22, 131–32, 153n17; gender and, 8–10, 13–17, 33–35, 58–59, 63, 94–98, 117–20, 125; generalizing, 79–80, 85–86, 100–102, 107–8, 115, 119–20, 130–31; knowledge and, 1–4, 14–15, 29, 34–35, 40–41, 63, 64, 86–92, 98–99, 122, 128–29, 131–32; maxims and, 1–17, 34–35, 40–41, 62, 64, 79–80, 85–92, 98–101, 105, 107–8, 115, 122, 125–26, 128–33; novel and, 5–7, 12–17, 19–20, 33–35, 58–59, 62, 68–70, 74–75, 79–80, 85–90, 100–101, 115, 117–20, 125–31; particularized, 3–7, 16–17, 62, 91–92, 94–95, 97–98; private, 85–86, 88–89, 126–28, 132–33; science and, 1–10, 19–20, 29–30, 40, 128–29
experimentalism, 3–8, 19–20, 27–30, 56–57, 66–67, 89–92, 128–29, 145n4

feminism, 37–38, 40–41, 109–10, 155n34
Ferguson, Frances, 13–14, 35, 74, 94–97
form: gender and, 51–52, 57–58; maxim as, 27, 33–34, 49–53, 57–58, 61–63, 65–69, 79–80, 87–89, 91–92, 97–99, 105, 120–21, 132–33; novel and, 16–18, 90–91, 94–97
François, Anne-Lise, 115, 165n31
free indirect discourse, 80–82, 104, 110–12, 115, 123, 126–27, 156n42

Gallagher, Catherine, 89–90
gender: class and, 104–7, 116–17, 131; experience and, 8–10, 13–17, 33–35, 58–59, 63, 94–98, 117–20, 125; form and, 51–52, 57–58; ignorance and, 108, 110–12, 114–20, 123, 132–33; interest and, 56–59; interiority and, 33, 75–76, 110–14, 120–21; knowledge and, 34–35, 39, 41, 44–45, 55–57, 62–63, 92, 110–14, 116–20, 123, 132–33, 165n31; maxims and, 8–10, 13–15, 39–40, 49–53, 57–58, 87–88, 92, 104–6, 108–10, 112–13, 115, 132–33, 164n20; novel and, 106–7, 109–20, 131–32; pedagogy and, 105–7, 109–14, 116, 123, 129–30, 132; science and, 31, 37–38, 44–45, 55–57
generalization, 12–14, 35, 51–52, 58, 85–89, 91–93, 95–102, 115, 119–20, 125–33
Giguère, Francine, 50–51, 137n23
Girten, Kristin, 3–4, 30–31, 37–38, 56–57, 142n82, 143n83
gothic fiction, 103, 117–20
Greenblatt, Stephen, 87

Habermas, Jürgen, 75–76, 155n34
Hacking, Ian, 91–92
Hadot, Pierre, 46, 78–79

Hanlon, Aaron, 17–18, 117–19
Harrison, William, 7–8
Hershinow, Stephanie Insley, 33, 165n31
historical fiction, 48–49, 58–59
Hooke, Robert, 16–17
Hoyle, John, 42
humility, 5–7, 27–31, 34–35, 45–48, 52–57, 63, 67–68, 132–33

ignorance, 5–10, 12–15, 18, 33–34, 40, 73–74, 108, 110–12, 114–20, 122, 125, 132–33. *See also* knowledge
individualism, 12–13, 33, 65–66, 69–70, 161n43
induction, 3–9, 11, 18–23, 27–30, 70–72, 76, 89–92, 136n7, 141n68. *See also* Bacon, Francis
inference, 3–4, 17–19, 22–23, 89–90, 128–29, 140nn55–56, 158n14
interest, 33, 40–41, 47–48, 53–61, 78–79, 145n4, 149n51
interiority (inwardness): gender and, 33, 75–76, 110–14, 120–21; knowledge and, 5–13, 29–34, 36–38, 61–62, 70–71, 74–77, 79–80, 86–87, 89, 92–96, 102, 133–34; maxims and, 7–14, 29, 31–36, 38, 58–59, 61–62, 65–69, 74–75, 78–82, 84–89, 91–92, 106, 120–21, 155n30; morality and, 82–83; novel and, 5–7, 10–14, 29, 31–36, 65–66, 68–70, 74–83, 84–89, 91–97, 106, 110–15, 120–21, 132–34, 163n7, 167n11; privacy and, 31–32, 35, 75–77, 82–83, 84–86, 88–89, 94–96; realism and, 10–11, 13–14, 29, 35, 68–70, 75–76, 87, 89, 91–97, 102, 115, 132; science and, 5–7, 21–22, 27–33, 35, 75–77, 155n30

Johnson, Claudia, 119–20
Johnson, Samuel, 89–92; *Dictionary of the English Language*, 16
Jost, Jacob Sider, 98–99

knowledge: broken, 11–16, 34–35, 37–38, 41, 67, 70–71, 82, 91–92, 128–29, 138n28; class and, 92–93, 131; as collaborative, 8–10, 22–23, 44–45; experience and, 1–4, 14–15, 29, 34–35, 40–41, 63, 64, 86–92, 98–99, 122, 128–29, 131–32; gender and, 34–35, 39, 41, 44–45, 55–57, 62–63, 92, 110–14, 116–20, 123, 132–33, 165n31; ignorance and, 5–7, 10–11, 33–34, 73–74, 110–12, 114–20, 122–23, 131; interest and, 53–57, 149n51; interiority and, 5–13, 29–34, 36–38, 61–62, 70–71, 74–77, 79–80, 86–87, 89, 92–96, 102, 133–34; of Love, 37–40, 57, 60–61; maxims and, 1–16, 25–29, 34–41, 47–49, 54–55, 61–62, 64–65, 67–68, 73–74, 79–80, 86–92, 97–99, 105, 122, 125–26, 129–31; method and, 20–29; novel and, 3–8, 10–11, 13–18, 29, 33–36, 69, 79–80, 86–90, 94–99, 116–20, 123, 125–26, 133–34, 159n24; as partial, incomplete, or inadequate, 1–3, 5–11, 17–18, 21–29, 31–33, 44–45, 65, 67–68, 70–71, 79–80, 89, 92–94, 123, 128–31, 133–34; privacy and, 76–77, 94–96; realism and, 86–87, 89–98, 102, 131; as relational, 4, 63, 90–91, 102; science and, 3–5, 8–10, 18–19, 21–29, 32–33, 56–57, 60–61, 64–67, 70–71, 89, 97–98, 128–29, 131; unknowing, 37–38, 40–41, 55, 60–61, 73–74, 91–92, 117–19, 165n31
Knox, Vicesimus, 106–7
Kramnick, Jonathan, 69–70

La Chapelle-Bessé, Henri de, 45–46, 146n9
Lafayette, Madame de: *La Princesse de Clèves*, 48–49, 58–59; *Zayde*, 48
La Rochefoucauld, François de, 10–11, 128–29; *Réflexions ou Sentences et Maximes morales* (*Moral Reflections or Sententiae and Maxims*), 8–10, 34–35, 37–42, 44–62, 132–33, 144n1, 145n6, 146n9, 146n15, 150nn65–66, 151n69

Latour, Bruno, 93–94, 158n10
Lee, Wendy Anne, 95–96
Lennox, Charlotte, *The Female Quixote*, 117–19
liability, 33
Licensing Act, 76–77
Locke, John, 12–13, 19–20, 29, 60–61, 82–83, 90–91, 106; *An Essay Concerning Human Understanding*, 8–10, 31, 69–70; *Second Treatise of Government*, 37
logic, 18–19, 22–23, 31, 44–45
Love, 37–43, 50–53, 55–63. See also knowledge: of Love
Lucretius, *De rerum natura*, 55–56, 149nn45–46
Lukács, Georg, 29, 92–93
Lynch, Deidre Shauna, 96

Macpherson, Sandra, 29, 95–96, 101–2
Maioli, Roger, 16–17
Marlowe, Christopher, *The Jew of Malta*, 87
Marriott, Thomas, 109
materialism, 37–38, 40–41, 55–56, 96–97, 102, 132–33
maxims: defining, 1–3, 16; experience and, 1–17, 34–35, 40–41, 62, 64, 79–80, 85–92, 98–101, 105, 107–8, 115, 122, 125–26, 128–33; form and, 33–34, 38–40, 49–53, 57–58, 61–63, 65–69, 79–80, 87–92, 97–99, 105, 120–21, 132–33; gender and, 8–10, 13–15, 39–40, 49–53, 57–58, 87–88, 92, 104–6, 108–10, 112–13, 115, 132–33, 164n20; as generalizable, 87–92, 95–102, 115, 125–30, 132–33; ignorance and, 5–10, 12–15, 40, 108, 122, 125, 131–33; interest and, 53–55, 59–61, 78–79; interiority and, 7–14, 29, 31–36, 38, 58–59, 61–62, 65–69, 74–75, 78–82, 84–89, 91–92, 106, 120–21, 155n30; knowledge and, 1–16, 25–29, 34–41, 47–49, 54–55, 61–62, 64–65, 67–68, 73–74, 79–80, 86–92, 97–99, 105, 122, 125–26, 129–31; as method, 42–45; morality and, 6–10, 14–16, 33–35, 44–49, 52–53, 68–69, 78–81, 85–87, 89–90, 105–6, 108–10, 128–29; novel and, 1–8, 10–11, 13–17, 29, 33–36, 48, 66–67, 78–82, 84–90, 92, 95–102, 106, 108, 120–21, 125–26, 128–31, 133–34, 159n23; pedagogy and, 109–10, 129–31; rationality and, 9–10, 33–34, 47–48, 64, 68–69, 78–80; realism and, 10–11, 13–14, 16–18, 29, 33–35, 68–70, 86–94, 97–98, 100–102, 125–26, 128, 130–31, 159n23; science and, 5–10, 18–22, 31, 34–35, 37–40, 44–45, 70–71, 106, 130; as self-canceling, 1–3, 10–11, 94–95, 131–33; virtue and, 45–49, 53–54
McKeon, Michael, 81–82, 90–92
metaphor, 21–22, 24–27, 32–33, 75–76, 153n17
Miller, D. A., 121–22, 163n7, 167n11
mimesis, 16–18, 33–34, 68, 90–91, 95–96. See also realism
Montagu, Mary Wortley, 87–88, 100–101
morality: ignorance and, 6–7, 14–15, 117–19; interiority and, 80–83; knowledge and, 6–10, 14–15, 56–57, 68–69; maxims and, 6–10, 14–16, 33–35, 44–49, 52–53, 68–69, 78–81, 85–87, 89–90, 105–6, 108–10, 128–29; novel and, 14–15, 33–35, 68–69, 80–83
More, Hannah, 105–7
Moretti, Franco, 29

new materialism, 93–94
novel: empiricism and, 16–17, 19–20, 69–70; experience and, 5–7, 12–17, 19–20, 33–35, 58–59, 62, 68–70, 74–75, 79–80, 85–90, 100–101, 115, 117–20, 125–31; form and, 16–18, 33–34, 90–91, 94–97; gender and, 106–7, 109–20, 131–32; generalization and, 85–86, 89, 91–93, 95–102, 115, 119–20, 125–27, 130–33; histories of, 3–5, 12–13, 16–18,

INDEX 183

33, 65–66, 69, 91–92, 117; ignorance and, 12–15, 18, 33–34, 115–20, 125, 131; interiority and, 5–7, 10–14, 29, 31–36, 65–66, 68–70, 74–83, 84–89, 91–97, 106, 110–15, 120–21, 132–34, 163n7, 167n11; knowledge and, 3–8, 10–11, 13–18, 29, 33–36, 69, 79–80, 86–90, 94–99, 116–20, 123, 125–26, 133–34, 159n24; marriage and, 122, 125; maxims and, 1–8, 10–11, 13–17, 29, 33–36, 48, 66–67, 78–82, 84–90, 92, 95–102, 106, 108, 120–21, 125–26, 128–31, 133–34, 159n23; pedagogy and, 109–14, 116, 130–31; privacy and, 77, 94–96, 126–28; realism and, 13–14, 17–20, 29, 33–35, 75–76, 86–87, 89–95, 97–98, 100–102, 115, 125–28, 130–31, 159n23; science and, 3–7, 17–18, 131

Paige, Nicholas, 48–49, 58–59
Pasanek, Brad, 5–7, 31–32
Paulson, Ronald, 74
plain style, 16–17, 69–70, 94–95, 153n17
Pope, Alexander, 8–10
Price, Leah, 105–7
Primer, Irwin, 42, 51–53
privacy, 31–32, 35, 60–63, 68, 75–77, 82–83, 84–86, 88–89, 94–95, 126–28, 144n3
probation, 22–29, 31, 37–38, 40–41, 55, 61–62, 65, 91–92
Proctor, Robert, 5

quixotic novels, 117–19

Raff, Sarah, 58
rape, 94–97
rationality, 9–10, 17–19, 22–23, 25, 33–34, 42, 46–48, 64, 68–69, 78–80, 109–10, 128–29, 153n12
Rawson, Claude, 80–81
realism: experience and, 87–89, 130–31; formal, 33–34, 69–70, 89; generalization and, 96–97, 100–102, 125–27, 130–31; individualism and, 69–70;

interiority and, 10–11, 13–14, 29, 35, 68–70, 75–76, 87, 89, 91–97, 102, 115, 132; knowledge and, 19–20, 86–87, 89–98, 102, 131; maxims and, 10–11, 13–14, 16–18, 29, 33–35, 68–70, 86–94, 97–98, 100–102, 125–26, 128, 130–31, 159n23; novel and, 13–14, 17–20, 29, 33–35, 75–76, 86–87, 89–95, 97–98, 100–102, 115, 125–28, 130–31, 159n23
Richardson, Samuel, 1–2, 69–70, 75–76, 108, 131–32; *Clarissa*, 10–11, 13–14, 35, 84–102, 106, 112–13, 132–33; *A Collection of Such of the Moral and Instructive Sentiments, Contained in the Preceding History, As are Presumed to be of General Use and Service*, 86–87, 157n2; *Pamela*, 106; *Sir Charles Grandison*, 87–88
Robert, Marthe, 15–16
Rogers, G. A. J., 19–22
romance, 33–34, 39. *See also* amatory fiction
Royal Society, 3–7, 19–20, 29–30, 56–57

Sablé, Madame de, 44–45
satire, 2–3, 10–11, 29–32, 35, 38–39, 64–69, 71–77, 80–82, 85–86, 95–96, 128–29, 151n11, 153n12
Savile, George, Marquis of Halifax, 109
Schaffer, Simon, 3–4, 19–20, 37, 56–57, 90–92, 96–97, 135n6, 145n4
Schiebinger, Londa, 5
Scholasticism, 21–22, 24–25, 31–32, 37, 44–45, 65–66, 92–93, 140n56
Schoolmen. *See* Scholasticism
Schwartz, Alexandra, 127–28
science: experience and, 1–10, 19–20, 29–30, 40, 128–29; gender and, 31, 37–38, 44–45, 55–57; ignorance and, 5–10, 18; interiority and, 5–7, 21–22, 27–33, 35, 66–67, 75–77, 155n30; knowledge and, 3–5, 8–10, 18–19, 21–29, 32–33, 56–57, 60–61, 64–67, 70–71, 89, 97–98, 128–29, 131; maxims and, 5–10, 18–22, 31, 34–35, 37–40, 44–45,

science (*continued*)
  70–71, 106, 130; method and, 20–29; modesty and, 3–4, 7, 10–11, 19–20, 37, 40–41, 44, 56–57, 89–90, 96–97, 145n4, 147n19, 149n47; novel and, 3–7, 17–18, 131; witnessing and, 3–4, 7, 10–11, 37–38, 40–41, 56–57, 86–87, 89–90, 128–29, 136n7, 145n4
secrecy, 60–61, 63, 144n3. *See also* privacy
secret history, 39, 48–49, 58–61, 85–86
Segrais, Jean Regnault de, 48
sententiae, 35, 68–69. *See also* maxims
Shapin, Steven, 3–4, 19–20, 37, 56–57, 90–92, 96–97, 135n6, 145n4
Snider, Alvin, 55–56, 149n46
Sprat, Thomas, 5–7, 29–30
Staves, Susan, 62–63
Steele, Kathryn, 101
Stoicism, 34–35, 44–48, 78–79, 147n16
Swift, Jonathan, 1–2, 7–8, 10–11, 31–33, 82–83, 85–86, 129–32; *The Battel of the Books,* 31–32; *Gulliver's Travels,* 2–5, 8–10, 14–15, 35, 64–69, 78–82, 157n48; *A Tale of a Tub,* 35, 65–69, 71–77, 79–80, 92–93, 106, 153n12, 157n48
syllogism, 22–23

Thompson, Helen, 3–4, 90–92

Valincour, Jean-Baptiste, 48–49, 58–59
virtue, 45–49, 53–54

Warner, William, 86–87, 97–98
Wats, Gilbert, 23–24
Watt, Ian, 27, 33–34, 69–70, 75–76, 89, 91–92, 94–95. *See also* realism
Wehling, Peter, 18
witnessing, 86–87, 89–90, 136n7, 145n4; modest, 10–11, 37, 40–41, 56–57, 89, 128–29, 149n47; sensitive, 37–38; virtual, 3–4, 7, 37, 135n6

www.ingramcontent.com/pod-product-compliance
Lightning Source LLC
Chambersburg PA
CBHW032100230426
43662CB00035B/856